Peter Damaske

Acoustics and Hearing

With 58 Figures

Dr. Peter Damaske
Zeppenlinstr. 21
38640 Goslar
Germany

Library of Congress Control Number: 2008922551

ISBN 978-3-540-78227-8 Springer Berlin Heidelberg New York

Springer is a part of Springer Science+Business Media.

springer.com

Typesetting: Data converted by VTEX using a Springer LaTeX macro
Cover: WMX Design GmbH, Heidelberg

Printed on acid-free paper SPIN 12044610 57/3180/vtex 5 4 3 2 1 0

Part I

Head-Related Sound from Two Loudspeakers

Preface and General Introduction

In a good concert hall, a big orchestra can evoke remarkably spacious sounds. The concertgoer is surrounded by the physical sound waves, and out of these waves the subjective sound impressions are created ***in the listener's head***. This biological "measuring device" organized itself during childhood. With no special effort—and quite continuously—it performs parallel data processing and makes no distinction between complex or simple analysis.

Imagine that the orchestra, after complete silence, strikes a load, abrupt chord. In that moment we can watch the immediate response of the hall. The phenomenon is called the *onset of reverberation*. During this process the listener's auditory system has to evaluate the *direct sound* of each particular musical instrument along with portions of sound reflected from the walls or the ceiling. If those sound reflections arrive at the listener's ears less than 50 ms later than the direct sound, they are called *early reflections*. All these amounts of sound make up such an intricate mixture that the ear is unable to resolve it as a series of separate events. From a favourable seat in the auditorium the listener receives only one complex impression, which can be wide and yet very detailed and appears abruptly in the front. This subjective impression may briefly be named a *sound image*. Its width and its depth, its facets and the weights or contrasts of its different parts characterize "the acoustics" of the concert hall and also the orchestra.

The onset of reverberation is followed immediately by a slow decay process, in which we can also perceive directional and spatial effects. This works very well after a distinct flourish of the orchestra when clouds of sound, floating over the listener's head, can be perceived. The duration and timbre of the reverberation and our impressions of width and depth are substantial for the acoustical quality of a room. As mentioned, this holds just the same for the onset of reverberation and depends on the music presented. However, in the decaying reverberation no directional details can be distinguished because the sound is now mixed stochastically after many reflections.

For this short description of the sound phenomena we started with orchestral music. In fact, when judging a concert hall we have to consider the style of music and even its separate passages because the acoustical quality of a room depends on these aspects. In a good hall we may be surprised by extremely nice glitters of sound in just one striking part of a particular piece of music, even in a single chord.

Exciting sounds often give rise to discussions of acoustical issues just *because* subjective impression comprises a fan-like variety of topics. However, when using physical methods to investigate the acoustics of a concert hall, we dismantle the maze of interconnected problems as much as possible and then clear each question step by step.

Indeed, we do not need the full orchestra for our investigations. For example, the orchestra may be replaced by the simple sound of a "peaceful" shot fired on the stage. A microphone, placed at a selected seat in the hall, receives the *impulse response* which is recorded for later analysis. A non-physicist might think it rather strange to replace an orchestra by a pistol, thus ignoring the richness of acoustical problems. Yet, if applied by experienced persons and with restraint, the sophisticated methods and devices of physics are excellent tools. Spectral analysis and the oscilloscope are typical examples.

However, powerful measuring techniques should not be allowed to mislead us to forget that acoustical quality is basically defined by *subjective sound impressions*. Moreover, data gathered by technical equipment can be so complex that their relevance to the listening experience will be unclear. The aforementioned impulse response of a room is a very good example: How is it to be interpreted, since it contains the complete, intertwined acoustical information? We know, for example, that the early reflections do substantially determine the sound image, whereas many of the later reflections are *inaudible*. Their unimportance is thus out of the question: they are redundant [1]. This makes the proper interpretation of measured physical data a difficult job.

On the other hand, direct comparison tests with many subjects are a time-consuming effort right from the start. Such experiments demand careful preparation and skilful execution in order to allow meaningful and reliable conclusions. All physical parameters have to be kept constant during the experiments, possibly for several months in extreme cases. This is a good reason for further preferring purely physical measuring techniques in room acoustics.

The foregoing discussion shows that relations between the data measured by technical devices on the one hand and subjective sound effects on the other should be analyzed carefully. Do we know sufficiently well, for example, how the sound image changes if one of the important early reflections is made stronger by 2 dB, or if it arrives 4 ms earlier at the listener's ears? Could we reduce the measuring redundancy right away with the use of technical models of characteristic properties of the human ear? In this book we shall devote our attention to questions of that kind.

Sound waves are subject to acoustic diffraction when they pass the human head with its funny-shaped pinnae. In this process the amplitudes as well as the phases of the waves are affected to a relevant degree. These changes depend on the frequency as well as the direction of the sound waves, and are of basic importance to our directional perception. However, where is the sepa- rating line between the external physical world and the subjective world of perception? Instead of seeking that line at the diaphragms of microphones placed somewhere in the sound field, we can better find it at the eardrums. We should take advantage of this fact and push the application of well-developed physical methods up to this limit.

Toward that end, the first chapter of this book describes a special stereophonic 2-channel system. The system uses the human head as an important acoustic object in the sound field. The idea is realized by a dummy-head that has two built-in microphones, allowing binaural recordings. The recordings are manipulated electronically and then reproduced by two loudspeakers in a living room. The result is a perfect surround transmission.

The second chapter deals with the onset of reverberation caused by the direct sound and some early reflections. The process can be simulated in an anechoic chamber. For a series of six early reflections we measure their respective *drift threshold*. This threshold is defined by a criterion which is sharper than that of the well-known Haas effect [2, 3]. If the sound pressure levels of all the successive reflections are adjusted to their drift threshold, the onset of reverberation has surprising properties. The resulting sound images are investigated with groups of subjects.

In the third chapter, we carefully examine these measured directional sound distributions. From those examinations, we arrive at a definition for a sound image's subjective *diffuseness*. Thus, the diffuseness can be characterized by a numerical value. A few detailed comparisons show that the known directional *diffusity* [4, 5], as measured with a unidirectional microphone, has hardly any meaning for the subjective impression of diffuseness. The auditory space perception requires binaural data processing which does not agree with the application of only one microphone.

The fourth chapter shows the drift thresholds to be computable with an astonishing accuracy. The calculations are realized by applying a loudness model described by A. Vogel and E. Zwicker [6, 7], and thus the close relations between the loudness and the drift thresholds become obvious.

Actually, time functions of the loudness even allow a quantitative distinction between the clear effect of the direct sound on the one hand, and the diffuse directional impressions created by some early reflections on the other hand. This discussion is presented in the fifth and last chapter of the book. It also reveals that the cross-correlation process involved in directional analysis might be combined with loudness analysis.

During the long reverberation a large multitude of reflections are received by the ear, and it tries hard to perform an analysis which is as precise and as fast as possible. The results presented here throw some light upon the permanent overstrain of the auditory system which, in reaction, creates diffuse and spacious sound impressions.

We necessarily repeat some information in the introductions to the chapters, so that each chapter is comprehensible in itself. The mutual relationship between the five chapters and thus the intention of the book is to emphasize the important part of the head and the auditory system in acoustics. The circle is closed by the head-related stereophony which allows the nice space effects—for example, the diffuse onset of reverberation—from concert halls to be replicated in a living room.

Acknowledgements

I am very grateful to the Drittes Physikalisches Institut at the University of Göttingen and its directors E. Meyer† and M.R. Schroeder for initiating the ideas presented here, ideas which I could never give up. In an animating cooperation with W. Burgtorf†, D. Gottlob†, V. Mellert, K.F. Siebrasse†, and B. Wagener we even managed to create echoes in the anechoic chamber. An intensive exchange of ideas with Y. Ando began in 1971, right from the first moment of his stay in Göttingen. It had a great scientific resonance in Japan.

The measurements described in this book, especially those with groups of subjects, were carried out in Göttingen up to 1972. Later on, some experiments and advancements were continued in my private home in Goslar. This work covers a period of many years, because my personal energy had to be concentrated on my teaching employment as master of a Gymnasium. The scientific results are published here for the first time. However, the contents were partly presented in a few papers given in Germany, Canada, and Japan.

I am also grateful to my wife, Christiane, who for several decades supported me in her friendly manner. She always followed the ideas and can sometimes explain some important points better than the author himself. Even the mess of cables in our living room did not cause grumbles, except after several months. I thank my children, Martin and Bettina, as well as Stefan Fricke for their stimulating criticism, and I wish to express my extra special thanks to Bettina again, who talked with so much patience to a silly dummy-head. I am thankful to the whole family for accepting so many subjects in our home, and I am equally thankful to all the persons in Göttingen and Goslar—might it be a few hundred?—who took part in the listening sessions. Finally, my thanks to Kenneth Allard for skilfully polishing the English translation.

Goslar, January 2008 *Peter Damaske*

Contents

Part I. Head-Related Sound from Two Loudspeakers

Part II. The Hearing Process in Concert Halls

1

Head-Related Stereophony

1.1 Introduction

The first stereophonic record of the world was probably edited in 1943 in the USA and contained Sousa's march, "The Stars and Stripes Forever." After the Second World War the stereophonic technique spread rapidly, though it requires at least two transmission channels. Its expansion was promoted by the first radio stations working with ultrashort waves, largely because they were partly arranged for stereophonic transmissions right from the start. The new technique was meant to allow spacious effects from the recording room to be perceived in the reproduction room as well. However, the complete and perfect transmission of all the spatial information could not be realized with two channels, although the human auditory system does not need more.

Spacious sound can be noticed in a mono-transmission as well because even a single loudspeaker will release the reverberation of the reproduction room itself. And with it the reverberation of the recording room is also audible. These effects are combined psychologically, and the overall spaciousness may be judged as satisfactory if there is little demand for quality. Supporting this opinion, a young lady cutting magnetic tapes at the Westdeutscher Rundfunk in Cologne never needed more than a single loudspeaker in a small portable radio, especially in her reverberating bathroom! In discussing stereophony we might keep this idea in the back of our mind.

When using two or more channels, pseudo-stereophonic effects will appear once again, or they may even be deliberately created as is often done in modern stereophony. Listening to symphonic music reproduced by a couple of good loudspeakers in a car while driving on the Autobahn at high speed can be very impressive, indeed.

In contrast to this pleasant idea, a group of terrifying dinosaurs pouncing on the poor cinemagoer—optically as well as acoustically—brings up the question whether the eye or the ear possesses more power to influence human feelings. So far, the edge goes to pseudo-stereophonic effects.

When the stereo technique was introduced after the Second World War, engineers strived for the ideal transmission. It could not be achieved right away, however, because the distorting effects were too prevalent.

In full contrast to monophony, let us now regard an extremely wide sound source—for example, the Concertgebouw Orchestra, Amsterdam with its 120 musicians, presenting "Till Eulenspiegels lustige Streiche" (Till Eulenspiegel's Merry Pranks), composed by R. Strauss [8]. According to a well-established seating plan, the musicians are placed at different points on the platform, and their instruments emit sound in different directions. The sound waves definitely spread all over the hall, especially after a few reflections off the walls or the ceiling.

Various methods are available for proper transmission of all the directional and spatial effects perceivable in a concert of that kind. Let us begin by describing a method with great technical requirements. The reader may imagine a hemispheric surface with a radius of two or three meters, domed over a seat in a concert hall. About a dozen directional microphones may be evenly distributed on this surface. They are meant to receive the sound waves arriving from the walls or the ceiling when they are passing the dome's surface. For each microphone a corresponding loudspeaker in the reproduction room is positioned on a surface of the same shape and size. All these loudspeakers cooperate in reproducing the sound waves received in the recording room. A listener sitting in the centre of this second hemisphere will probably have similar acoustical impressions as he or she would in the recording room, especially if the reproduction room has sound-absorbing walls.

The reader may decide to what extent this transmission method might be practicable. In any case, concert halls have already been simulated for scientific reasons through the use of 65 loudspeakers which were evenly distributed on a hemispheric surface in an anechoic chamber (see Fig. 2.1, page 47). Thus, for the first time, the acoustical qualities of different halls could be compared directly in subjective tests [9, 10].

A Greek goddess named "Acustica" might be able to carry out such direct comparisons, because she can jump abruptly from one hall to the next. However, human beings—incapable of tricks of that kind while the music is going on—must rely on the aid of modern electronic devices.

When simulating a concert hall in an anechoic chamber with 65 loudspeakers, we cannot reproduce signals which were "simply" recorded with 65 corresponding microphones. Instead, only a few specific sound reflections of the concert hall are modelled electronically along with its genuine reverberation. About ten loudspeakers are sufficient for this purpose. The proper delay times and the sound pressure levels of the modelled reflections can be adjusted according to the architectural plan of the room under investigation, or according to a measured impulse response of the room. For comparing different rooms all parameters must be switched simultaneously, which is no serious problem [9–11].

But back to stereophony: When 2-channel stereophony was introduced for the first time, the bar was not put very high. The electroacoustic imaging of a complete concert hall is a nice job nowadays! At first it was intended only to transmit an image of the orchestra itself, including only partially the onset of reverberation caused by the stage and the backstage area.

This can be achieved with the use of only two microphones, positioned in front of the orchestra at a distance of a few meters. In order to reproduce the sound in a living room, two loudspeakers are required, one each placed at the left and the right in front of the listener. In a transmission of that kind, the particular musical instruments are perceived at their proper place, roughly speaking. The space between the loudspeakers is filled by phantom sources, as they are called, created by the data processing in the brain. Thus, we perceive sound in a part of the living room which is not caused by real sound waves travelling in that region. In the best case, the acoustical images of the instruments even have their proper distances. The reverberation of the recording room is also audible, but as in monophony it is superposed by the reverberation of the reproduction room. By and large, we have a stereophonic image resembling the real concert hall in many directional and spatial details. Therefore recordings with two microphones might be regarded as contenting, sufficient, and rather simple.

The stereophonic image in the living room is decisively determined by the recording technique. Quite often, when recording an orchestra, numerous microphones are applied for the soloists or groups of instruments.

With the use of this extensive multichannel recording, a sound image for 2-channel reproduction may be composed later on and a good balance can be struck between the groups of instruments. However, there are other techniques to employ when using only two microphones. For example, we can place a separating circular disc between the microphones [12]. In the following, two standard methods are discussed for reference, but only the reproduction with loudspeakers will be regarded.

1.2 Two Standard Stereophonic Methods

In a transmission intended to work by *intensity differences*, we use two unidirectional microphones placed at about the height of the sound source. They are fixed quite close together—for example, one on top of the other—and their axes are usually twisted by more than ±45°, related to the main sound direction. For a transmission using *time differences*, however, omnidirectional microphones are more suitable. Their mutual distance is 40 cm, for example, and their connecting line is perpendicular to the 0°-direction.

The left part of Fig. 1.1 is valid for two directional microphones L/R of the super cardioid type, with their axes at ±50°. For sound arriving from +30°, for example, the sensitivities of the microphones differ by about 8 dB, as indicated by two small circles in the diagram. However, a level difference of more than 12–15 dB is required for the directional impression in the reproduction room to reach +30°, which is the loudspeaker position. In order to overcome this problem, the microphone axes might be twisted by more than ±50°, or the distance between the microphones and the orchestra might be reduced, thus increasing the recording angle (at least for some of the musical instruments).

A microphone arrangement for a transmission using time differences is shown in the right part of Fig. 1.1. If a large recording distance is desired, the mutual distance of the microphones may simply be increased. A pair of microphones arranged in this way may be seen in concert halls, often placed above the stage front.

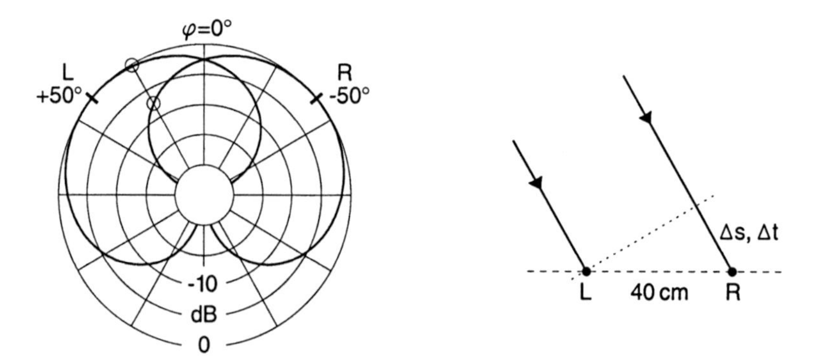

Fig. 1.1. Microphone arrangements for intensity or time differences

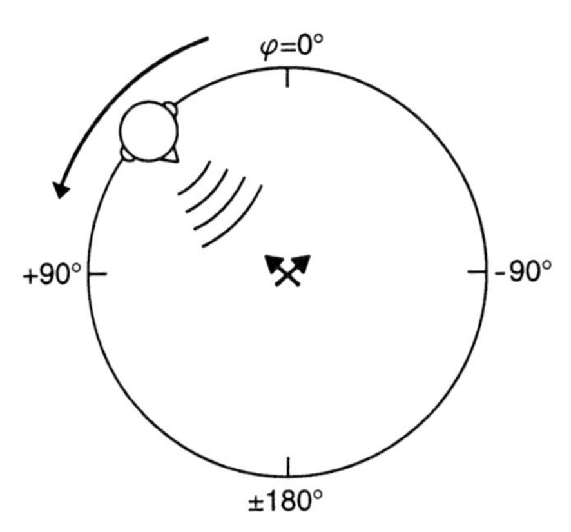

Fig. 1.2. Recording scheme with two microphones, talker at azimuthal angle φ

When judging a stereophonic transmission, an important question is whether the directions are properly reproduced or not. The discussion of this problem shall be related to the standard arrangement of the two loudspeakers in the reproduction room. Thus, they are positioned at the $\pm 30°$ angles in respect to the listener's line of view.

To what extent do the directions perceived in the reproduction room equal the physical sound directions in the recording room? The following experiment is commonly used to answer that question: In the recording room, a circle of about 2 m radius is defined; the circle includes an angular scale. At the height of a talker's head two microphones—for example, with cardioid characteristics—are installed at the centre (see Fig. 1.2).

The talker stands on the circle line and utters a short phrase toward the centre of the circle. If he or she talks from the 0°-direction, the listener in the reproduction room should hear the speech in the 0°-direction too, that is, from the front. As is known, this works rather precisely. Now, the talker chooses new positions again and again, each time repeating the phrase toward the centre of the circle. If all the directions were reproduced correctly, the direction perceived by the listener would equal the real direction of the talker each time. This is indicated by the dashed line in Fig. 1.3. However, all directions actually perceived remain in the frontal range between the loudspeakers. The reverberation of the reproduction room might eventually be audible and disturbing, but we want it to be disregarded. Furthermore, the voice may appear at a certain elevation angle, but this is not experienced as a serious problem with the loudspeakers positioned at $\pm 30°$.

The diagram is now explained in detail. If, in the recording room, the talker stands in the angular range from $-30°$ to $+30°$, the perceived direction

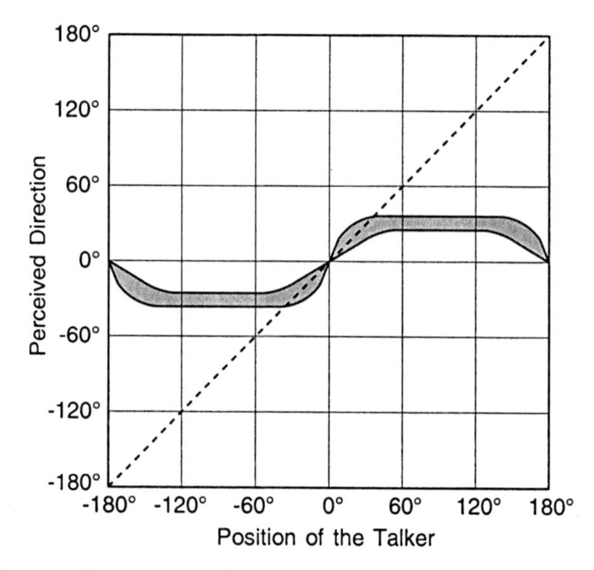

Fig. 1.3. Stereo transmission by intensity or time differences

equals the recording direction (shown in the central region of the diagram). However, for the large range from 30° to 150° in the recording room (on the left), the speech is constantly perceived near the left loudspeaker (at about 30°). In the diagram this is indicated by a stripe that is parallel to the abscissa. If the talker starts at 150° and moves toward the direction 180° (to the rear), the perceived direction, beginning at the loudspeaker, moves back to 0° (to the front). Equivalent details are valid for the right side of the head. The grey region is meant to indicate that the transitions between different parts of the curves are fluent and that several different curves are possible.

Details depend on the recording method, the properties of the microphones, and the positions of the microphones. An experienced sound engineer will be able to adapt the central part of the transmission curve to the width of the sound source. That is to say, he or she will manage the musicians to appear properly placed just in the frontal region of the image. In a transmission based on time differences, some sound portions may eventually be perceived a little beyond the loudspeakers. In view of all these possibilities it becomes clear that the curves in Fig. 1.3 are meant as a general explanation, not as the result of one specific measurement.

Finally, we have to consider the user of the stereo equipment: If the listener decides to place the loudspeakers at ±20° instead of ±30°, the overall sound image will be too narrow. However, extremely large loudspeaker angles are also a problem. In this case, disagreeable elevation angles might occur for the important central parts of the image, but we do not want the voice of a talker floating in the upper bookshelves!

To summarize the description presented in Fig. 1.3, regarding the standard 2-channel transmissions we can say that the perceived directions cover a range of about 60° in front of the listener. As already mentioned this limitation is not as important as suggested by the pure numerical value. However, another imaging problem cannot be neglected and is not represented by the curves in the diagram: transmissions having a high directional resolution often show a lack of "depth". In this case the musical instruments of an orchestra may appear in a sort of layer which can be perceived as a thin curtain of sound hanging in front of the listener. This means that the distances are not properly reproduced.

It is interesting to confront the two standard stereophonic methods by comparing the subjective effects caused by *wide-band noises*. In order to simplify the listening procedure, pink noise was copied from a compact disc [13] to the hard disc of a computer [14]. Thus, noise samples of 3 s duration could be sequenced with pauses of 1 s duration. The sound pressure at the listener's seat was 75 dB(A), and all samples were adjusted by ear to the same loudness.

When generating directional impressions by *intensity differences*, the following details can be noticed (left part of Fig. 1.4). When both loudspeakers are generating the same sound pressure level, the noise appears centrally in the front, having distinct bassy components. If the two sound pressure levels gradually deviate from equilibrium, the noise moves sideways in the horizontal plane. The directional impressions stay precise; the perceived distance equals the loudspeaker distance; the sound colour changes (not shown in Fig. 1.4); and, in the main, the bass dominance decreases toward the sides. When the sound pressure levels differ by than about 20 dB, the sound is perceived in the direction of the loudspeaker. When increasing the imbalance,

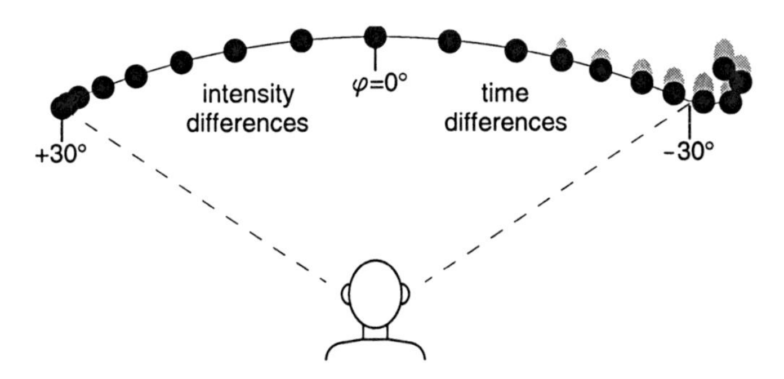

Fig. 1.4. Perceived directions with pink noise, constant loudness

the perceived direction does not move beyond the loudspeaker direction. The sketch is meant as a slanted top view from behind the listener and is based on the author's personal impressions.

In the other fundamental method, stereophonic impressions are generated by *time differences* between the two signals (right part of Fig. 1.4). However, at 0 ms time difference the sound impression is just the same as it was at 0 dB level difference in the preceding case. This is because in both cases the loudspeakers are operated with exactly the same signals. With a growing time difference between the two signals the perceived direction moves sideways again. From about 0.6 ms the sound is perceived in the direction of the loudspeaker. Up to this value the sound colour is nearly constant. In some directions, narrowed images are noticeable, partly with upward extensions. For time differences above 0.6 ms, the perceived direction moves beyond the direction of the loudspeaker by up to 5°(estimated). From about 0.8 ms the image moves back to the loudspeaker and expands upwards.

With test records, the reader may carry out such comparisons with noise signals, but unequal loudspeakers (crossover networks) or too much reverberation may cause serious confusion. For music recordings the described observations have the following consequences.

Stereo images with soft edges can be realized in transmissions based on time differences. This is especially welcome for orchestra recordings. However, if the distance between the microphones is too large, the time differences for lateral sound may partly exceed 700 μs, which is about the limit in natural hearing, and the sound image may have sharp edges.

That problem does not occur when using intensity differences, but in this case, too, the image can have sharp edges. Those edges indicate that large amounts of sound received from the lateral range in the recording room are monophonically reproduced by one loudspeaker. In general this recording method allows a somewhat more precise localization of musical instruments in the image, but the image seems less spacious.

Can we perhaps combine the two standard methods in order to avoid some of their problems? The reader may imagine a pair of directional microphones having a horizontal distance of about 20 cm from each other, which is the distance of our ears, and the microphone axes being a little spread. Wouldn't this setup amount to a simple model of the head? As the answer seems to be positive the next mental step will certainly be the idea of making a dummy head.

But back to the problems of the standard stereo transmissions! The following calculations show that the common reproduction with two loudspeakers causes *spectral colorations* of the sound. In order to explain this fairly unno-

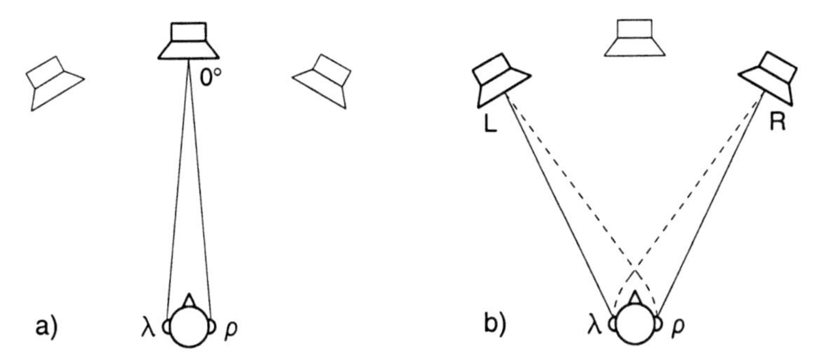

Fig. 1.5. Two symmetric situations; crosstalk and channel mishmash in (**b**)

ticed problem, *three* loudspeakers are considered; these are positioned at the directions ±30° and at 0° in the front (Fig. 1.5). All loudspeakers may have the same distance of 3 m from the listener.

If only the frontal loudspeaker is working, or if the two loudspeakers at ±30° are sending exactly equal signals L/R, a central image will appear in the front. However, equal loudnesses presumed, spectral differences are perceivable if we switch over from one of these possibilities to the other. The differences are explained in the following with reverberation neglected.

If only the 0°-loudspeaker is working, the levels at the ears λ, ρ are equal, symmetry of the head presumed. Let us now compare the two signal paths $L \rightarrow \lambda$ and $L \rightarrow \rho$ in the right part of Fig. 1.5. The frequency may be adjusted to 1 kHz, for example, and the right loudspeaker may be switched off. In this case, the sound pressure level at the left ear λ is increased by 2.4 dB, and decreased by 3.7 dB at the right ear ρ, as compared to frontal sound incidence (amplitude values 1.32 and 0.65). This amplification or attenuation results from sound diffraction at the head. Figure 1.6 shows the general relations for a wide frequency range. The error margins enclose three results each, taken from the literature, and the circles show data points measured with a dummy head [15–17].

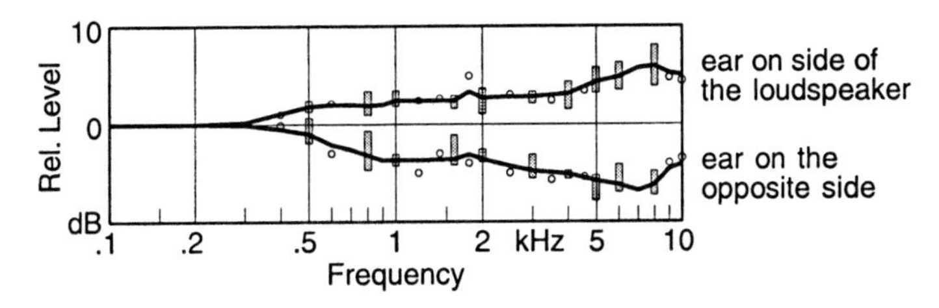

Fig. 1.6. Sound pressure levels at the ears, ±30° angle of incidence

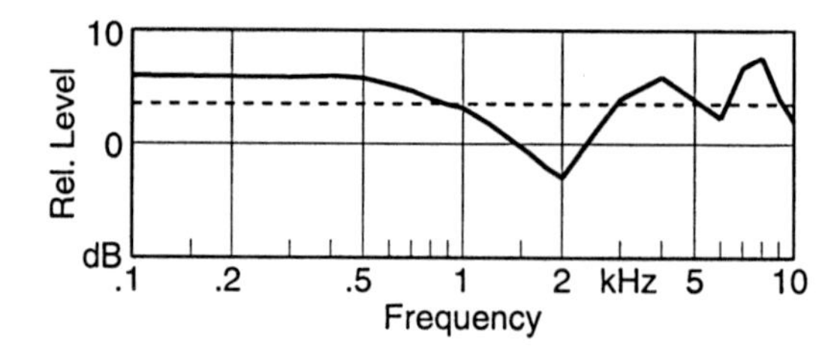

Fig. 1.7. Sound pressure at the ears for $L + R$, related to a 0°-loudspeaker

When both loudspeakers L, R are working simultaneously, their waves interfere at each ear. The weaker wave from the opposite side arrives about 260 μs later. At 1 kHz, for example, this results in a phase difference of 93.6° and for the amplitudes mentioned earlier the vector sum is $1.43 = 3.1$ dB. The general result for this interference phenomenon at the head is shown in Fig. 1.7 in relation to a singular 0°-loudspeaker.

A *comb-filter effect* is noticeable and causes spectral coloration. The dashed reference line stands for the singular 0°-loudspeaker. It might be expected that the signal level for this loudspeaker has to be set to +6 dB to cause equal loudnesses, but +3.5 dB is sufficient. In a living room the interference is disturbed by reverberation. Thus the comb-filter effect is partly concealed and obviously not even noticed when enjoying stereophonic music. This phenomenon reminds us of the commonly known effect of a piano pedal. Unfortunately, the annoying comb-filter effect is especially distinct with excellent loudspeakers.

1.3 Surround Sound in a Living Room, Pre-experiment

The discussion in the previous chapter reveals that the sound waves are considerably disturbed by the geometrical shape of the human head. In stereophonic transmissions it would be inconsistent to ignore this fact on the one hand, and to claim extreme technical precision of microphones, records, amplifiers, and loudspeakers on the other hand.

To reduce acoustical problems right away, a *dummy head* may be used for recordings of speech or music, and also for scientific measurements. This device is a replica of a human head with two built-in microphones. Careful modelling of the external ears is especially important and must include the construction of the eardrums and proper coupling of the microphones to the ears.

In the acoustical near field of the dummy head, interaural time and intensity differences do occur; these depend on the frequency and on the recording direction. The differences are caused by the geometrical form of the head and the ears. The cavities in the external ears are Helmhotz resonators which are excited differently, in accordance with the direction of the incident waves [17]. Stereophonic transmissions can take advantage of these details because the auditory system's data processing is based just on this natural encodement of directional information [18]. If regarded in this light the use of common microphones means giving away indispensable amounts of information.

With the use of a dummy head and its two microphones, and using the smallest possible number of transmission channels, an excellent reproduction of all the directions around the head was achieved for the very first time in 1969 [16, 19–21]. In the experiments carried out at that time, a large anechoic chamber was available (see Fig. 2.1, page 47). Later on it was interesting to find out if the head-related stereophony would work as well when using two loudspeakers in a typical living room. In the search for an answer to this question, a pre-experiment was carried out in 1978. It had a positive result. Following several technical advancements—with many intermediate tests and measurements over a period of several years—a main experiment was prepared. The result of the main experiment is published in this book. However, as an introduction, the pre-experiment shall be explained first. Most of the details in the experimental proceedings are valid for the main experiment, too, which is described in Sect. 1.4.

In the pre-experiment a dummy head constructed in 1969 was used again. It works with dynamical microphones that are connected to the ear canal by long probe tubes with an external diameter of 8 mm [22]. The "eardrums" are realized by thin circular slices of felt mounted in the ear canal. A natural timbre of speech is achieved in the accompanying pre-amplifier by adjusting the overall frequency response. The dummy head is covered by a felt wig roughly imitating a thick shock of hair. The wig is meant to amplify the spectral difference between the sound received from the front or the rear. To roughly model the upper part of a person's body, a jacket is fixed to the tripod below the neck of the dummy head.

When producing speech recordings the dummy head was placed in a wide corridor next to an inside staircase of a private home. The tripod for the dummy head was positioned in the centre of a circular piece of PVC carpet specially prepared for these recordings (Fig. 1.8). A woman talker was standing at the edge of the carpet, her mouth at the height of the dummy head's ears, always at a distance of about 1 m. The recording directions were indicated by 20 marks plotted on the carpet. With the use of these marks the talker could easily and precisely find the defined positions. As shown in Fig. 1.8 the

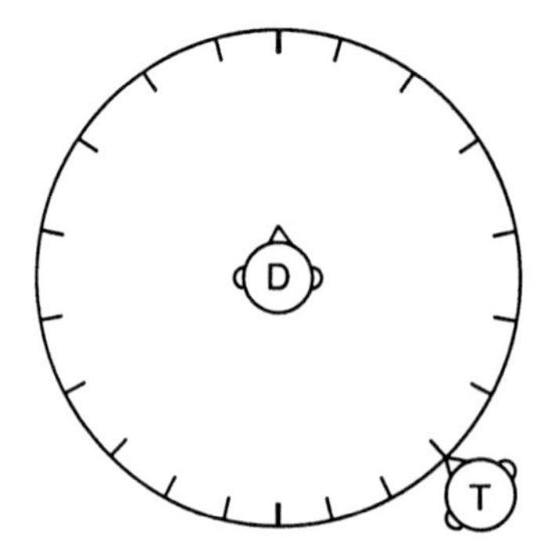

Fig. 1.8. Talking to the dummy head, 20 fixed recording directions

angular steps between the positions differed on purpose, with smaller steps in the front and in the rear. At a tape speed of 38 cm/s the following foolish sentence was recorded once and only once per direction: "Haben Sie tatsächlich am achtundzwanzigsten Februar schon einmal Artischockensalat gegessen?" (Have you really ever eaten artichoke salad on the 28th of February?)

Rather soon the woman talker was able to repeat the phrase again and again in a friendly intonation, always adjusting her voice to the same loudness and the same talking style. Using the set of 20 fixed positions shown in Fig. 1.8, the recording directions were selected in a quasi-stochastic manner. Neither steps nor other noises were recorded, thus avoiding anything extraneous to the talker's voice. If the reverberation of the recording room is very strong the subjects will have difficulty specifying directions. Therefore, a few woollen blankets were hung in the large corridor, which extends over three floors, in order to reduce its reverberation time to a value of about 1 s.

The tape recording was reproduced with two loudspeakers in an ordinary living room (Fig. 1.9). The room is nearly rectangular and has a size of 25 m^2. In one corner there is a passage to a veranda separated by a curtain. One of the walls is covered by bookshelves filled with books. The parquet floor is partly covered with carpets; the furniture, the curtains, and the windows are distributed with no symmetry. The reverberation time of this room is a little less than 0.4 s. The listener's seat is an armchair placed roughly at the centre of the room, at a distance of 2.70 m from each of the loudspeakers. As usual, and as already mentioned, the loudspeakers are positioned at angles of $\pm 30°$, related to the subject's line of view, which is defined to have the azimuthal angle $\varphi = 0°$.

The loudspeaker boxes were specially developed for these experiments, but they can be used quite generally. A detailed description is given in Sect. 1.6.1, but some information is required at this point. The wooden boxes

Fig. 1.9. Loudspeakers in bookshelves of the living room

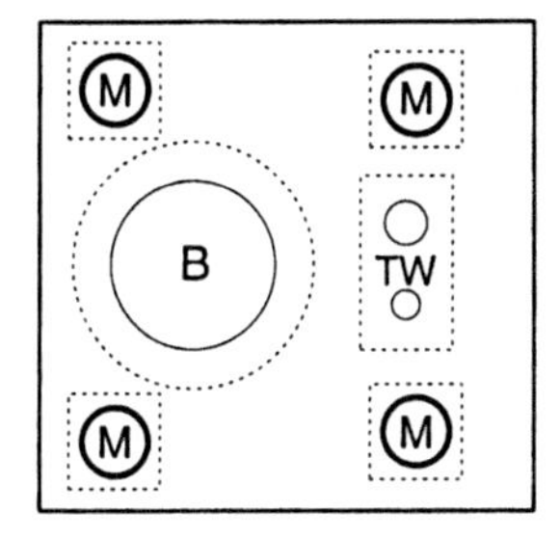

Fig. 1.10. Sound panel of left loudspeaker box, four middle range speakers

fit into a bookshelf and they are at the height of a sitting person's ears (Fig. 1.9). In each box four middle range speakers work in phase (Fig. 1.10). Their clear impulse response is a big advantage. It is achieved through the use of membrane diameters of only 7 cm, by silicon suspensions, and balancing of manufacturing errors within the group. However, the most important point is the directivity of the group. On the main axis of the group, the four partial waves do interfere positively at all frequencies. If the main axis is directed towards the listener's seat the incoherent reverberation of the living room is of less consequence than with conventional omnidirectional loudspeakers. When the stereo signals L and R arrive at the two ears λ and ρ they will thus be more precise and can generate clearer sound images.

However, the transmission system explained here is aimed at a higher target: The signals must reach the two ears *as separated as possible*, in short $L \rightarrow \lambda$ and $R \rightarrow \rho$. This could be easily achieved with headphones, but then the stereo image would be distorted in its most important range—the front. The “frontal” sound may be perceived at a considerable elevation angle, or it can appear quite near the head, occasionally even *inside* the head. Some methods for counteracting the problem do exist, but many find wearing headphones to be tedious anyhow. In this book we will keep to the reproduction by two loudspeakers.

When only one of the loudspeakers is working the sound will, unfortunately, reach both of the listener’s ears, and arrive on the “wrong” side as well. This effect is called “acoustical crosstalk”, and as already explained it causes directional distortions and spectral coloration in common stereophony.

The undesirable crosstalk can be counteracted by additional compensating signals. The principle was applied in earlier experiments concerning head-related stereophony [20]: The cross-talking sound from the left loudspeaker, for example, arrives at the right ear as well. It is extinguished by a small part of the left signal which has to be inverted and appropriately delayed, and which is emitted by the right loudspeaker. When preparing the pre-experiment a special electronic filter, described by the author, was reconstructed [21]. In this filter the compensating signals have to be generated and coupled to the other channel, respectively. In Sect. 1.6.2 the electronic filter is explained in detail.

With the described experimental setup the pre-experiment was carried out in 1978, and the dummy head recording of speech was reproduced in the living room. Altogether 120 subjects took part in the course of several months. Each person was led into the room separately and had to take a seat on the armchair immediately. In a bookshelf in front of the subject a circular scheme showing 20 numbered directions was clearly visible, the circle subdivided into equal parts. The subject was asked to imagine his seat to be placed in the centre of the circle. Then the listening task was briefly explained in the following words: “You will now hear a recorded voice speaking a single phrase, but you don’t have to follow up the contents of the phrase because it is rather nonsense. The sentence will be repeated several times, with sufficiently long intervals. During these intervals, please indicate the direction from which you heard the voice.”

In order to explain the circular scheme the experimenter went to the frontal 0°-position and asked: “From which direction do you hear my voice?”. In most cases the subject answered "number 10" immediately and correctly, otherwise a further explanation was necessary. In that case the experimenter went to the 90°-position on the right and asked for the number again (number 5). If the subject hesitated, or if the answer was again incorrect the experimenter asked from another direction in the range at the right. If the subject asked what to do if a direction between two numbers appeared, the experimenter answered in a comforting way: "In that case, just take one of the next numbers. Your answer isn't so very important since many persons take part in the experiment." None of the subjects tried to discuss or to analyze the acoustical space perception. Finally the subject had to position his head properly. This could be achieved within about two seconds by optically matching the edge of a cupboard with a plotted line on the wall. The subject was asked not to move his head during the test and then the tape was started and reproduced with natural loudness. The recorder was placed in another room in order to avoid distraction.

The perceived direction could be indicated within an interval of 4 s prepared on the tape. The answers were noted by the experimenter, quietly sitting on a sofa behind the subject. When the test was finished after four minutes many subjects asked immediately: "Now, where do you have all your loudspeakers?". Searching in the room they could find only two loudspeakers in the front, appearing quite normal (Fig. 1.9). *117 out of 120 subjects spontaneously perceived the speech from behind if it was recorded from behind.* The subjects and the experimenter himself were equally surprised by this fact.

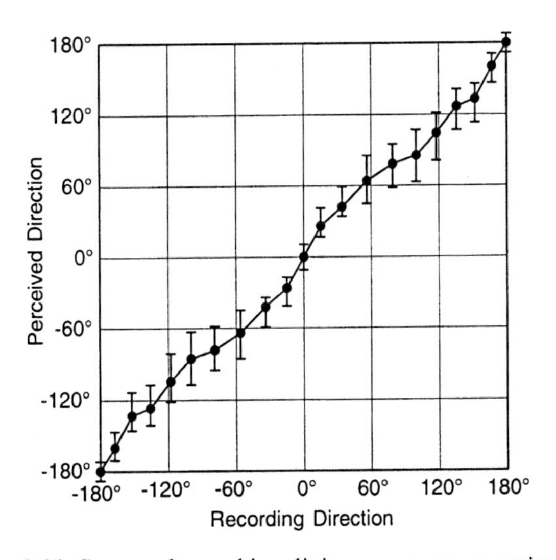

Fig. 1.11. Surround sound in a living room, pre-experiment

The listening results are plotted in Fig. 1.11. The diagram shows that the perceived directions are quite like the recording directions all around the head. For each recording direction the answers are distributed near a central value which is plotted as a point. The central 50% of the distribution is enclosed by the error margins shown for each point. By the way, the diagonal line cannot be expected to be the result of this experiment, because even with natural sound in the acoustical free field the results are similar to Fig. 1.11 [23].

For each pair of symmetrical recording angles, for example, $\pm 15°$, $\pm 34°$, $\pm 56°$, ... the distributions of the perceived directions from both sides of the head were compared. In each case the two distributions belonged to the same statistical ensemble, with a certainty of 95%. For this reason the results from both sides of the head were combined before plotting them in Fig. 1.4. However, we should not forget that the transmission works in a 360° range all around the head. Some of the subjects were asked if they had noticed elevation angles of the perceived voice. The answers were not taken down but showed that the speech was essentially perceived in the horizontal plane.

Subjects of all age groups between 13 and 68 years took part in this pre-experiment. Of the persons who agreed to take part *no one was excluded*. All their results are included in the diagram. Five subjects knew their task in advance, but it was probably unknown to all the others. When leaving the room each subject was requested not to talk about the contents of the experiment.

1.4 Surround Sound in a Living Room, Main Experiment

The pre-experiment showed that surround sound can be realized with only two loudspeakers in a living room. After this proof the precision of the technical equipment had to be adapted to the standard of modern digital devices. This was carried out step by step and supervised in short listening sessions. No part of the equipment remained unchanged. Finally, the experiment of a head-related stereo transmission was repeated with many subjects.

The first enhancement of the devices was to employ precise condensor microphones in a newly designed dummy head [24]. In comparison with the dynamical microphones used in the pre-experiment, the new microphones provide a better signal-to-noise ratio, a wider frequency range, and an increased precision of the sound image. Second, a third dummy head was made with better mechanical stability and enhanced symmetry of the external ears (see Sect. 1.6.3).

As mentioned earlier, the sound waves from each loudspeaker reach the ear on the opposite ("wrong") side of the head as well. The electronic filter that compensated for this acoustical crosstalk was also improved step by step. It is now a well-defined printed circuit in a small metal box (Sect. 1.6.2).

Last but not least the loudspeaker boxes had to be reconsidered and rebuilt in three versions (see Sect. 1.6.1). In comparison with the pre-experiment, their directivity was somewhat diminished. It is now just sufficient for the surround effect. Even without the aforementioned compensation, the directivity of the loudspeakers causes very clear sound images when reproducing any common music recording. The frequency response was flattened and extended to lower frequencies. In such experiments, the two boxes should be as equal as possible, especially with respect to the crossover networks, as phase errors would gravely damage the sound image.

All changes to the equipment had to be judged by listening to many different records—orchestra music, solo instruments, singing, and speech. This was especially true with the test records [13]. When two or more alternatives were compared by switching over, no cracks could be heard and the loudness did not change. An automatic repetition of CD samples was possible. The following working steps were carried out more or less cyclically:

- judging the sound image: geometrical properties, clarity, naturalness;
- judging the surround effect with dummy head recordings;
- changes in the loudspeaker boxes: spectrum, symmetry, directivity;
- changes in the circuit wiring of the electronic compensation filter.

The dummy head recording for the main experiment was carried out on noise-reduced tape with a velocity of 38 cm/s [25]. This analog recording was digitized with a resolution of 16 bits, and then stored and digitally processed in the computer (ADD). With the application of only a few specially programmed procedures, the overall quality could be clearly enhanced: weak transitions between the phrases and the intervals, and full noise elimination during the intervals. The so-called DA converter is also installed in the computer [14].

Finally the surround effect was checked by a group of 56 subjects aged between 16 and 60 years. Most of the subjects were recruited from courses at a nearby adult education centre, which caters to a wide variety of students from all over Germany. Having all kinds of jobs, they could not have any knowledge about the experiment. The measuring procedure and the evaluation do not differ from the pre-experiment described earlier, except for a shorter spo-

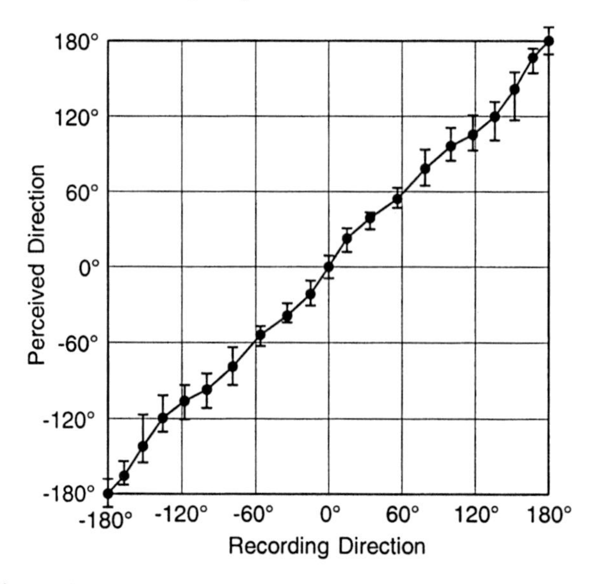

Fig. 1.12. Surround sound in a living room, main experiment

ken phrase that was used this time: "Hat niemand kürzlich im Monat September Kastanienblätter gesucht?" (Has anybody picked up chestnut leaves recently in September?).

The curve in Fig. 1.12 shows that the surround effect is maintained on the same level when using the precise technical equipment. However, two restrictions have to be mentioned: If it seemed to be advantageous the channels were interchanged when sequencing the phrases in the computer. Furthermore, on comparing two symmetrical recording directions on both sides of the head, the perceived directions occasionally differed strongly. In these cases the "better" response was sometimes selected for the evaluation. This concerns seven of 56 subjects for the directions $\pm 158°/\pm 133°$, and three subjects for two other pairs of directions. (A few months after the termination of this experiment the loudspeaker boxes were once more considerably improved in the low frequency range. This was achieved with two simple strips of wood.) Let us summarize the result: A true high-fidelity transmission was realized, and now included all directional and spatial aspects.

1.5 Experiences with Common Stereo Recordings

Not only dummy head recordings but *any common stereo recording* can be adequately reproduced with the system described. This compatibility is very important in order to increase the general acceptance of the ideas. In the sound image of an orchestra, a pop band, or a soloist the following details can be observed when the compensation is *switched on*, and were quite pronounced in the moment of switching over:

- no loudness difference, no disagreeable spectral changes;
- no unnatural elevation angles of musical instruments or voices;
- widening of the image, increase of the stereophonic depth;
- directions and distances of all instruments/voices became more natural;
- high image stability when the head is moved sideways ($\approx \pm 25$ cm);
- no abrupt quality loss when moving forwards/ backwards ($\approx \pm 25$ cm).

One more detail has to be especially emphasized. On judging the sound of single musical instruments, including the piano, a certain spectral change can be observed at the moment of switching on the compensation: the effect might be called a *decoloration*. (The reader may remember the disturbing comb-filter effect discussed in context of common stereo reproductions, Fig. 1.7). The decoloration by compensation can also be observed with speech or singing. It is astonishing how different and how natural a voice will be suddenly experienced when it is no longer spoiled by a comb-filter! The same relief can be experienced with pop music. It is worth mentioning that the author does not remember any impairment of any common recording by the compensation effect. The aforementioned positive aspects are particularly clear when replaying dummy head recordings. However, good loudspeakers are indispensable.

These observations, probably described a little euphorically, lead to the conjecture that directional precision and precise distances in the subjective sound image are intrinsically linked with spectral purity.

1.6 Description of the Equipment

A perfect spatial transmission was the main thing to be realized when designing the loudspeakers, the compensation filter, and the dummy head. Another matter of considerable interest was improved reproduction of common recordings. Old records and modern compact discs made with the usual microphones exist in many styles and huge numbers; these are a great cultural treasure. Fortunately, the two lines of interest mentioned turned out to coincide, at least with respect to reproduction. In the following the technical equipment will be described only in its final state, as used in the main experiment. However, the process of equalizing the frequency characteristic of the dummy head is described in more detail, as it allows for some far-reaching conclusions.

1.6.1 The Loudspeakers

In a large living room the loudspeaker boxes may be placed at a certain distance from the walls; this is advantageous for stereophonic reproduction. In smaller rooms, however, and if the boxes are simply put into a bookshelf, the sound image will usually be less perfect. Nonetheless, when the experiments described were carried out in a living room of only 25 m^2 (see Fig. 1.9) the loudspeakers were placed in bookshelves. The intention was to prove that even in this *disadvantageous* case the surround effect can be realized.

The boxes are made of 12 mm plywood reinforced internally. Their external dimensions are: width/height = 52 cm, depth = 32 cm. They just fit into a bookshelf each with a free space of 1 cm. Their front sides are put at a slight angle with respect to the rear wall; the outer edges jut out a little more than the inner edges.

The sound-transparent cloth covering is fixed by magnets. If it is removed the slanted sound panel is visible (Fig. 1.10). It consists of 13 mm chipboard and is internally reinforced. The 148 mm bass woofer is placed at half height. Next to it, towards the inner edge of the box, there are two tweeters one on top of the other. Four 70 mm middle range speakers are arranged approximately at the corners of a rectangle (w 32 cm/h 35 cm). Inside the boxes these speakers are separated from the woofer by domed plastic containers damped by porous material. This feature prevents any intermodulation. The sound panel is additionally stabilized by a strip of wood and by the seven loudspeaker chassis. Separated by a felt seal, the sound panel is screwed onto strips of wood which are attached to the walls of the box. The closed box is carefully damped with the use of glass wool. The crossover networks are mounted on the outside of the sound panel. Thus they are easily accessible, but invisible if the cloth covering is in place.

By adjusting the slant of the sound panel within the box, or by adjusting the box itself, the main sound beam can be directed toward the listener.

The directivity of the loudspeaker boxes (Fig. 1.13) was measured in a garden when it was covered with 10 cm of snow. Warble tones were used for these measurements. The centre frequencies of the third octave ranges are given in the polar diagrams plotted in relation to the listener's head.

At frequencies between 400 Hz and 4 kHz the main lobes have widths between 70° and 120° measured at −5 dB, if we neglect the splitting. Above 6.3 kHz the widths are between 50° and 70°, determined by the tweeters. As a result of the overall bundling the stochastic reverberation of the reproduction room will be less powerful at the listener's seat. Thus the direct sound com-

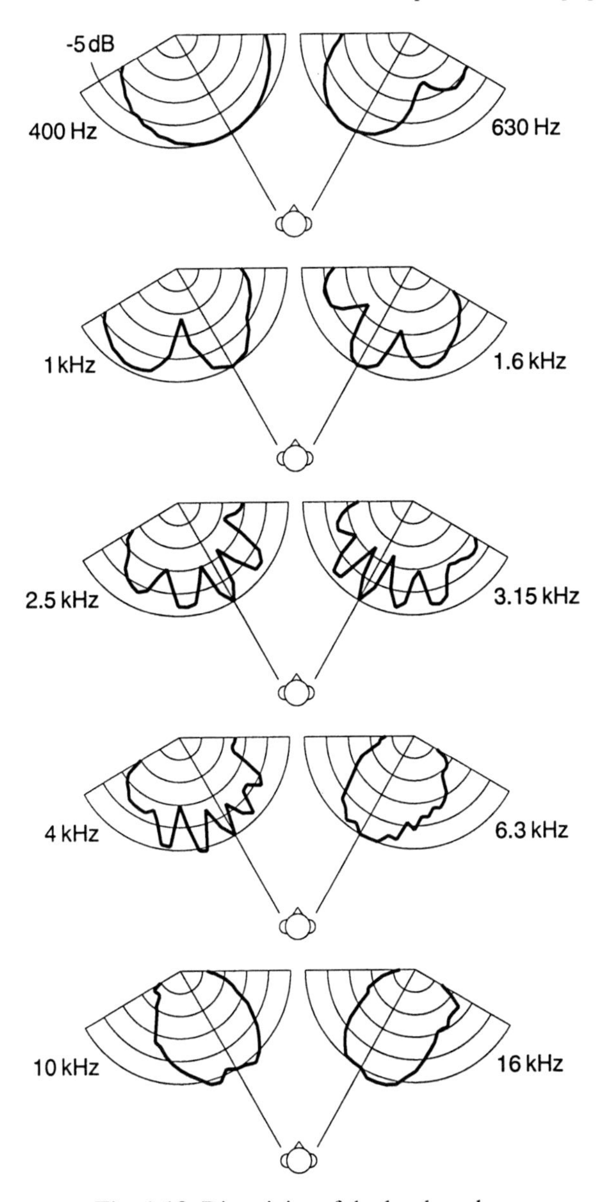

Fig. 1.13. Directivity of the loudspeakers

ponents can interfere more clearly at the ears. This results in greater precision of the sound image.

If the head is moved sideways the centre of the stereo image will follow this movement to a smaller extent than with omnidirectional loudspeakers. Let us explain this observation for the case in which the compensation is switched off: If the subject moves to the left, for example, the direct sound component of the left channel will increase less than without bundling, and the direct component of the right channel will decrease less. As an explanation we might imagine two tubes leading the sound from each loudspeaker towards the head. In a tube the sound pressure does not decrease with the length of its path. The main lobes of the loudspeakers are wide enough to keep the head inside when it is moved. Even at 4 kHz, which is at the upper limit of the frequency range suitable for the compensation, the small central lobe within the main lobe is wide enough to allow a lateral head movement of about ±25 cm.

When the head is moved forward or backward by ±25 cm no evident change can be noticed in the stereo image. This is quite comprehensible because the 0°-line is the axis of symmetry of the stereo setup. However, *even when moving sideways*, the image stability is remarkable if the directional loudspeakers described here are used. A probable reason is the mentioned constancy of the sound pressure levels at the ears. In any case interaural time differences do not depend on the directivity. The following details are noticeable when moving sideways:

- If the compensation is switched off, the central range of the image may follow the head movement, but this is not really annoying.
- If the compensation is switched on, the listener will be reminded of the holography when moving the head moderately. In this case the stereo image seems to stay at its place, just as in a real concert hall.

When constructing directional loudspeakers their frequency characteristic has to be equalized. The extent to which this could be achieved may be judged by looking at Fig. 1.14. During the measurement the loudspeakers were again placed in the bookshelves and then directed toward the listener's location (Fig. 1.13). A directional condensor microphone was positioned at this point in the room and at the height of the box centre [26]. The small deviations in

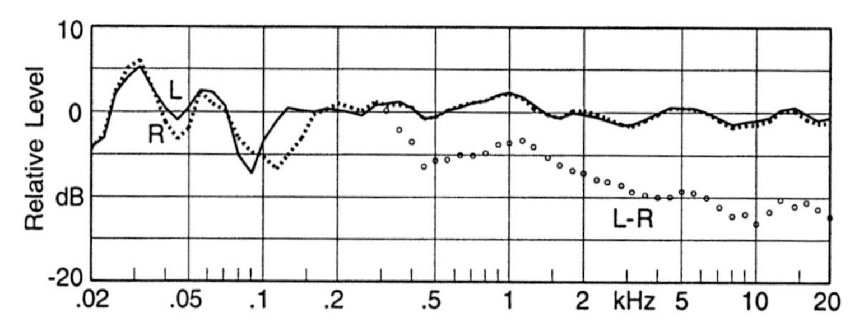

Fig. 1.14. Directional loudspeakers in a living room: f-response, compensation

its frequency characteristic were eliminated arithmetically. The main axis of the microphone pointed to the loudspeaker examined; thus the directivity of a human ear was at least roughly imitated.

The test signals consisted of sinusoidal tones. Their frequency was modulated in a range of $1/3$ of an octave. This corresponds to the "frequency groups" made up by the ear when processing acoustical data. However, juxtaposed points in the diagram are distanced by $1/6$ of an octave. Thus the details are shown unconcealed. The measuring accuracy is about ± 2 dB at 50 Hz and ± 0.3 dB at 1 kHz. Below 200 Hz the curves are impaired by box resonances and by resonances of the room sized 25 m^2. At frequencies near 130 Hz great differences between the curves L and R exist. However, a measurement in the acoustical free field showed two equal curves over the full frequency range.

At low frequencies definite resonances must be expected in any room. In a small room their disturbing effect may end only above 200–300 Hz. However, these resonances cannot seriously affect the stereophony since the human ear does not evaluate the three low octaves from 20 Hz to 160 Hz when generating directional impressions. Let us consider Fig. 1.14 once again: At all frequencies $f \geq 180$ Hz the two curves L/R do agree quite well, and in larger rooms they would agree even better.

The equality of the loudspeaker boxes is enormously important because it creates stereophonic precision. Therefore the equality was specifically checked in the living room, once again with warble tones. The channels L and R were operated at the same voltage, either one by one or simultaneously; and in the last-named case they were operated either in phase or out of phase. The microphone was placed at the listener's location and adjusted to the 0° direction. An interference test showed its position to be precisely on the axis of symmetry between the loudspeakers. The sound pressure level at the microphone is defined to be *0 dB if only one channel is working*, let us say L.

If the second channel is added in phase ($L + R$), the sound pressure level should be +6 dB. However, if the signal R is inverted ($L - R$), the result should be $-\infty$ dB if the sound waves completely cancel each other at the microphone. This ideal case is not realized in the living room, as shown by the open circles in Fig. 1.14. On the other hand positive levels of the difference signal $L - R$ are also imaginable (even +6 dB). The reasons might be faulty loudspeakers or acoustical problems of the room, but this case does not occur, especially not above 315 Hz. At 1 kHz, for example, the relative sound pressure level of the signal $L - R$ is not positive but about -6 dB. In the wide and important frequency range from 450 Hz to 5 kHz it is always 5–10 dB below the mono levels L or R (see Fig. 1.14). Compared with +6 dB, the value for the signal $L + R$—which virtually determines the stereo image—the gain is even 11–16 dB.

Thus the loudspeakers are equal to a large extent and they are generally suitable for creating stereophonic precision. Their high equality ensures that no irregular interferences arise when purposely using compensating sound components. Fig. 1.14 shows that this quality is discernible at frequencies above 315 Hz.

Some notes about the wiring scheme and the crossover networks have to be given. The electrical connections of each speaker chassis and, as already mentioned, all elements of the crossover networks are accessible on the front of the sound panel. The group of four middle range speakers covers the frequency range from 150 Hz to 5 kHz. The speakers have a nominal resistance of 4 Ω each. Forming two vertical lines, the speakers arranged on top of each other are wired in series, which makes a resistance of 8 Ω for each line. As the two lines are parallel the group of four again has an overall resistance of 4 Ω. This network has the advantage that production differences of the chassis are balanced out within the group. Furthermore, the central points of the lines were interconnected, thus forming a bridge and reducing the overall error even more. Theoretically the bridge current should be zero and therefore the connection was made by a wire. However, on listening to the effect of this measure, a weak coupling with a resistor of 4 Ω appears preferable. Nevertheless, the desired interference capacity of the group is caused mainly by its directivity, which reduces the influence of the quasi-stochastical reverberation.

In the crossover networks the bass woofers are faded out with a slope steepness of 12 dB/octave and the middle range speakers are faded in with 6 dB/octave. The two slope steepnesses between the middle range speakers and the tweeters are 18 dB/octave each. This transition had to be adapted to the given working limits of the chassis. Furthermore, a disagreeable resonance of the middle range speakers had to be diminished by the use of a rejection circuit.

Acoustical precision can be achieved only if the corresponding coils and capacitors in the two loudspeaker boxes are matched with extreme care. In order to guarantee good matching, Lissajou figures were examined on an oscilloscope. Phase differences are visible as elliptical deviations from a straight line. Circumspection is demanded, mainly at low frequencies, if this method is applied with the loudspeakers placed in the bookshelves: Geometrical irregularities of the room result in different acoustic loads on the individual speakers, thus modifying the phases at the test points of the crossover networks. (The Lissajou figure changes if a window pane is blocked by hand!)

In addition, the loudspeakers were tested with short, square pulses ($\leq$25 μs). By putting a microphone at the listener's location it is possible to check whether the pulses are stretched out in time. This might happen if the tweeters are carelessly placed on the sound panel, or because of phase problems in the tweeter network. It pays to ensure a proper pulse form, since this yields a precise and yet soft and natural sound of human voices or musical instruments.

1.6.2 The Compensation Filter

At low frequencies, it is not possible to compensate for cross-talking components. There are two reasons for this. First, compared with the wavelength, the head is small in this case, not making up a true obstacle. At 100 Hz and a wavelength of roughly 3 m, for example, the sound can bypass the head nearly undiminished (compare Fig. 1.6). Thus its components are nearly indistinguishable, and a detailed discussion would be pointless. Second, extreme or chaotic resonances of the reproduction room will appear at low frequencies (see Sect. 1.6.1). These two points may explain the widely known fact that low frequency components are unimportant for the acoustical perception of directions. In any case, a limiting frequency of 340 Hz was chosen for the electronic filter. The compensation shall be constructed only above that frequency. As depicted in the schematic diagram of the compensation filter (Fig. 1.15) the low-frequency components of the signals are carried directly to the loudspeakers.

The filter circuit is realized by analog techniques with the use of low noise operational amplifiers, mostly four in one case. The structure of the schematic diagram is symmetrical. Thus we need to follow up only one of the two input signals. We select the input L. As mentioned earlier the signal is split by a crossover network consisting of a high-pass and a low-pass filter. The two signal components are added again, and their sum is supplied to the output L'.

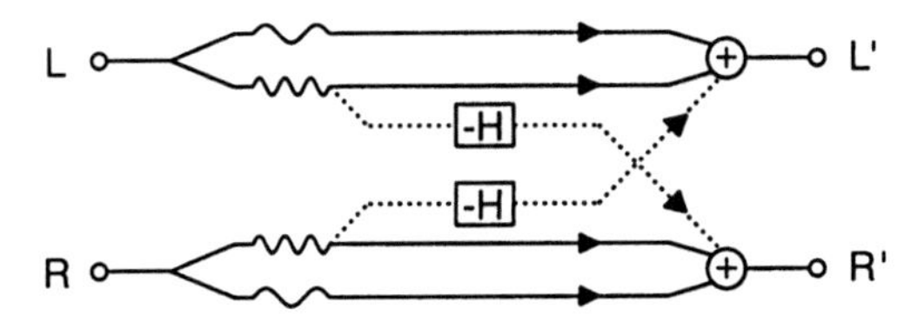

Fig. 1.15. Schematic diagram of the compensation filter

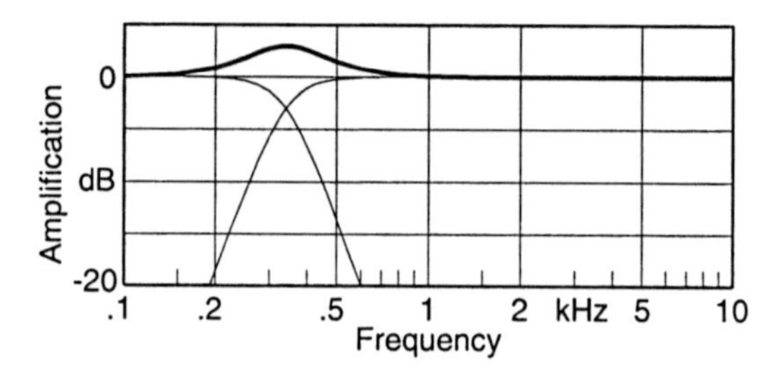

Fig. 1.16. Crossover network and sum-signal

The reason the signal has to be split and recombined will be considered later. Instead, we will discuss the problems that result from this procedure. If a sinusoidal voltage of the frequency f is applied to an analog filter, the phase of the output signal will be shifted by an angle $\varphi(f)$. In principle this might cause problems when the components are recombined. However, the filters actually realized are of the Butterworth type [27]. Since they both have the order four and equal cut-off frequencies, we have a simple phase relation. If $\varphi_{\mathrm{LP}}(f)$ is the phase shift caused by the low-pass filter, the equation $\varphi_{\mathrm{HP}}(f) = \varphi_{\mathrm{LP}}(f) + 2\pi$ is valid for the high-pass filter. However, the constant 2π makes no noticeable difference. Thus the signal components at the output of the compensation filter may be added without any phase problem.

This important advantage comes at a cost, however, which is expressed in Fig. 1.16. The frequency characteristics of the low-pass filter and the high-pass filter are plotted as two thin lines crossing each other at 340 Hz. The boldface curve is valid for the overall level which is increased in the transition range of the crossover network, having its maximum of 3 dB at 340 Hz. To correct this effect, the inverse frequency characteristic has to be realized by a band-stop filter.

In each channel the proper equalization is achieved by a single operational amplifier. It is arranged in the unbranched path right at the output and is not shown in Fig. 1.15. As long as the cross-connections of the circuit are inactive, the overall amplification is 1 at all frequencies and thus the circuit makes no sense. In this case, virtually no differences can be detected in speech or music when abruptly switching between the reproduction of the output or the input signals. Thus the phase distortions caused by the correcting band-stop amplifiers may be neglected, especially as they are equal in both channels.

Let us now imagine a person seated at the intended place in the living room. Once again we follow the signal L. From the left loudspeaker it arrives at the left ear, $L \rightarrow \lambda$, but because of the sound diffraction at the head it reaches the right ear as well, $L \rightarrow \rho$, and is slightly delayed. This crosstalk

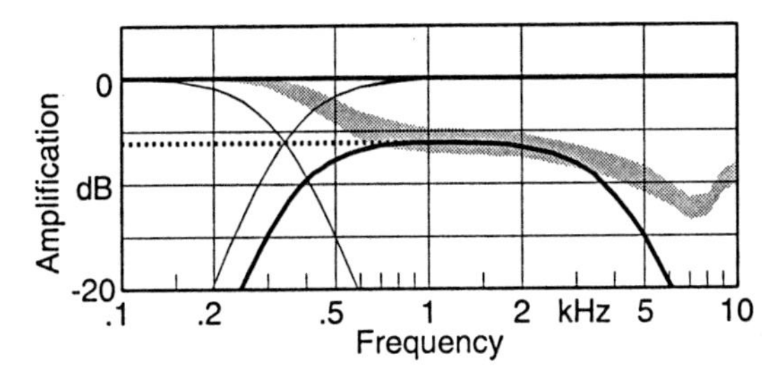

Fig. 1.17. Frequency dependence of the compensation level

is undesirable and must be compensated by sound from the right loudspeaker to the right ear: $R_{comp} \to \rho$. The compensating signal is constructed from the signal L by electronically simulating the sound diffraction at the head and inverting the result. In Fig. 1.15 this is shown by a dotted line and a box with the symbol −H. The construction of the filter −H requires three operational amplifiers. First we need a 3.3 kHz *low-pass* filter, and second the signal has to be delayed by an *all-pass* filter.

Figure 1.17 helps to explain the 3.3 kHz low-pass filter. The two thin lines are again valid for the crossover network at 340 Hz, but now the afore mentioned correction of the 3 dB maximum has been accounted for. Thus all curves are related to the straight 0 dB line, and the curves of the crossover network are somewhat lower in the transition range. They intersect at −6 dB instead of −3 dB.

The sound waves travelling from the left loudspeaker to the right ear, $L \to \rho$, have to bypass the head. Thus they are attenuated to an extent which depends on the frequency. The attenuation is shown by the lower curve in Fig. 1.6. However, when the sound from the right loudspeaker arrives at the right ear, $R \to \rho$, there is a pressure build-up near the head and the signal is amplified. This amplification is shown by the upper curve in Fig. 1.6. The difference between these two curves, including their inaccuracy, is represented by the grey strip in Fig. 1.17. It is this strip which determines the suitable level of the compensating signal.

The proper compensation level is achieved mainly by the 3.3 kHz low-pass filter. Its basic amplification of −6.1 dB is indicated by the dotted line in Fig. 1.17. This line passes into the boldface curve, which does not leave the grey strip between 700 Hz and 3.5 kHz. Thus the required compensation level is successfully maintained in that frequency range. At high frequencies the compensating signal fades out rapidly, and at 8 kHz its level is already 25 dB less than required. At high frequencies, however, compensation for the acoustical crosstalk would be a problem anyhow; and it would make no sense because our directional perception is based mainly on envelopes of complex signals [18]. This means that both the low-frequency range and the high-frequency components are more or less disregarded in directional analysis!

Toward *low* frequencies the boldface line in Fig. 1.17 follows neither the dotted line nor the grey strip, but it drops rapidly. Here, the compensation signal fades out automatically because it originates from the high-pass filter of the crossover network (Fig. 1.15). Thus, the boldface line tends to be parallel with the thin curve of the 340 Hz high-pass filter. At the intersection of the thin lines the compensation level is -12.1 dB. This value combines two data points: first, the -6.1 dB basic amplification of the 3.3 kHz low-pass filter simulating the head, and second, the -6 dB amplification of the 340 Hz high-pass filter. The slope steepness tends toward 24 dB/octave, and yet there are no phase problems. At any frequency the high-pass component has the phase shift $0 \bmod 2\pi$, related to each of the components of the signal R, and also to their sum. Let us summarize: In a wide frequency range, the amplitude of the compensation signal follows the sound diffraction at the head, but below or above that range the signal fades out. This terminates the discussion of the amplitude.

Hence, we can turn to the second problem: The sound propagation to the opposite side of the head causes a certain delay which has to be electronically simulated. Fortunately the delay time is basically independent of the frequency. This is known from wide-band correlation analysis [28]. Thus the required delay can be realized by a second-order all-pass filter [27]. This needs just one more operational amplifier per channel. With this simple solution a sufficient accuracy can be achieved, though only up to a 4 kHz limiting frequency. This is no real problem, however, because the compensation signal fades out anyway. If the low-pass filter which simulates the diffraction at the head is included, the constructed group-delay time is 235 ± 5 μs. This is close to the delay caused by the sound propagation to the ear on the opposite side.

As shown in Fig. 1.15 the complete compensation signal is inverted and then fed to the right channel. Thus it is subtracted from the signal R which is recombined from its low-pass and high-pass components. Apart from the minus sign and the aforementioned phase shift of 0 modulo 2π, and apart from the delay time purposely constructed, the output signal of the filter $-$H has no phase shift as compared with the signal R or its components. For this reason, when retransformed into sound, it can compensate or at least reduce the crosstalk from the other side of the head; and thus we have realized a better channel separation.

The consideration of the electronic compensation filter started from the input L but, of course, it is valid for the input R, too.

Let us briefly comment on the construction of the compensation filter. With the use of well-known electronic methods, the sound diffraction at the head was simulated rather precisely, but within certain frequency limits. The successful sound transmission all around the head proves that the construction is basically correct for a large group of subjects (Fig. 1.12). The described filter is different from earlier versions, but once again the compensating for cross-talking compensation signals was unnecessary.

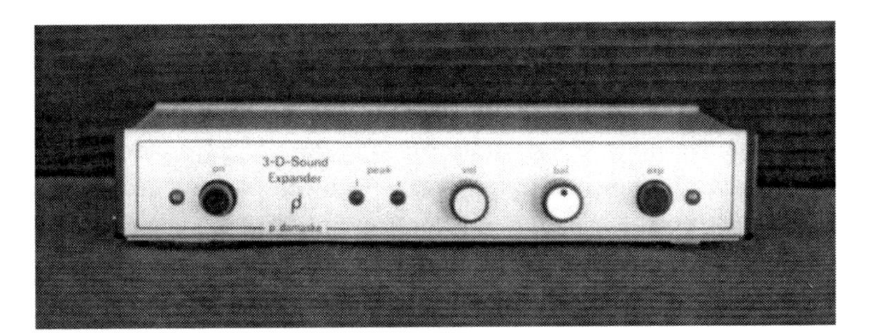

Fig. 1.18. The compensation filter

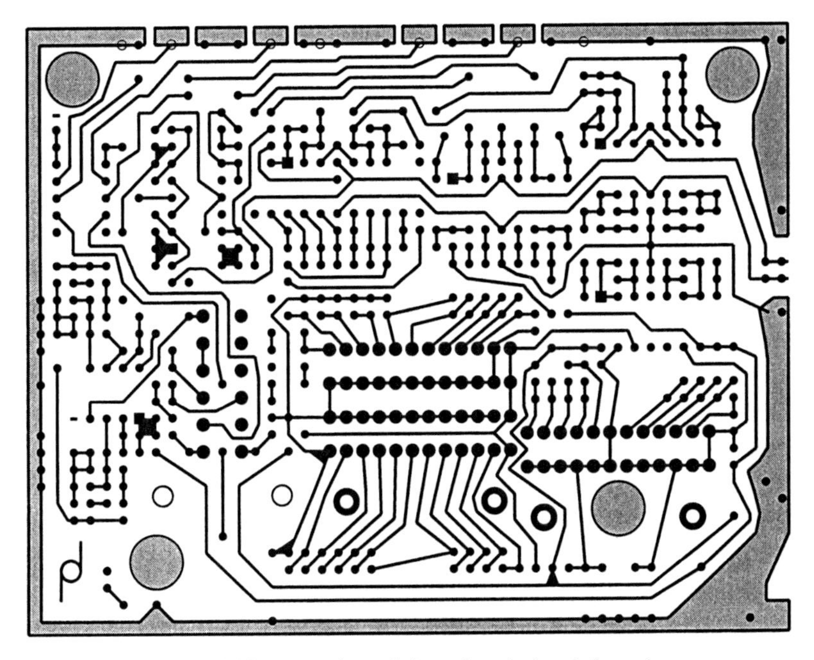

Fig. 1.19. A section of the printed circuit board

The following final remarks concern the technical implementation. The circuit is constructed using analog techniques and employs operational amplifiers. The circuit has a dynamic range of 110 dB. Its precision and its symmetry are determined by the electronic components. The resistors have a manufacturing tolerance of 1% or partly 0.1%. All capacitors have an accuracy of 1%. As known from radio technology, the signals $L + R$ and $L - R$ might help to avoid symmetry problems. However, this method was not applied, as the circuit should remain very clear during its development. Therefore the symmetry was again controlled through the use of Lissajou figures (see Sect. 1.6.1). The first operational amplifier in the circuit is combined with step switches. Thus amplifications from +15 dB to −10 dB can be chosen in steps of 5 dB, and the balance can be varied in steps of 0.25 dB. When the compensation is switched on/off there are no cracks or jump-like changes of loudness. In the chain of stereo devices the filter is introduced right at the input of the power amplifier. It has its own power supply, and if the power is switched off the signal path is not interrupted. The explanation of further details seems to be unnecessary. Figures 1.18 and 1.19 give a general impression of the device, showing the metal case and a section of the printed circuit board.

1.6.3 The Dummy Head

In an article from the 1940s, the author found a photograph showing a dummy head mounted on top of the head of a person using headphones. This somewhat funny arrangement had a medical purpose: It helped blind patients with a hearing defect to find their way or to follow a conversation in a group of people. Instead of outer ears, the dummy head in the photograph had flat microphones with a diameter of about 4 cm. It is clear from the photo that no front–back discrimination was possible with this device. (No scientific reference in this case.)

A wooden ellipsoidic dummy head, likewise with flat microphones and no outer ears, was used by E. Meyer and his students around 1938. In a lecture held in 1964 it was noted that the device did not work properly.

Fortunately, very small microphones existed in 1969 when the author was lucky to get the head of a display dummy from a department store in Göttingen [16]. With the use of this head, the basic idea was taken up once again. This time, however, the outer ears were not removed.

Let us now describe the dummy head used in the main experiment when recording speech from all around the head. This dummy head, Hugo #5, is shown in Fig. 1.20. Its detachable felt wig—imitating a thick shock of hair—

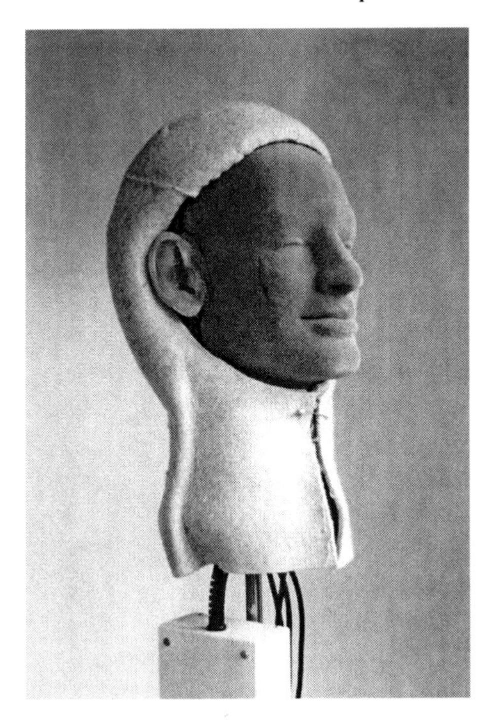

Fig. 1.20. The dummy head of the main experiment

was used only for the speech recording, but not for the measurements described later on in this chapter. The general effect of a wig on front–back discrimination seems worth investigating. However, apart from the documentation of frequency responses, the wig's effects require controlling the 360° stereo transmission with large groups of subjects.

Model heads made of rigid foam are commonly used to decorate shop windows. They are covered with felt to protect the basic material against damage. An average-sized model head of this type was chosen for the dummy head of the main experiment (Sect. 1.4). Epoxy resin was used to cast external ears of average size and shape. Burrs were filed off, and a hole for the ear canal had to be drilled. When attaching the ear casts to the dummy head, proper positioning was important. As a preparation for this work the head was sawn through in a plane behind the ears since the microphones and a mount had to be installed. Finally, the two half heads were reunited.

The mechanically and acoustically important juncture between the external ear and the microphone is shown in Fig. 1.21. The black areas indicate metal. A bent copper tube with an inner diameter of 5.9 mm begins at the

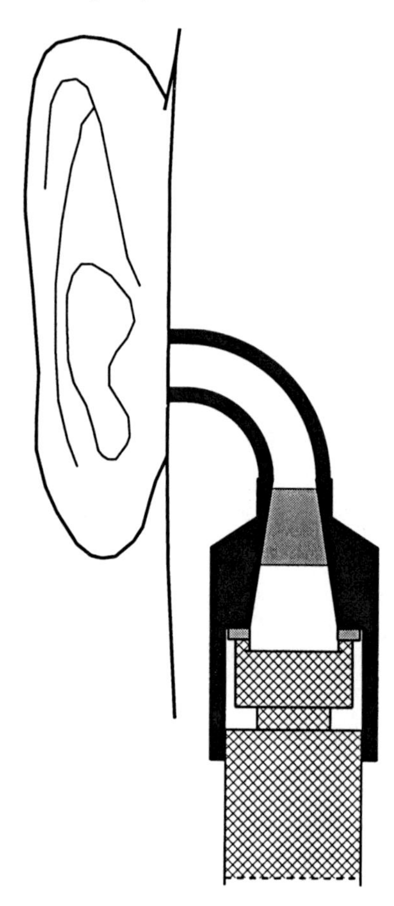

Fig. 1.21. Ear canal and juncture for the microphone, natural size

pinna and represents the ear canal. At the pinna it is encased by epoxy resin which makes a very stable junction. The tube leads to a brass juncture with a cylindrical hole into which the tube is soldered.

The inner space of the juncture opens conically up to a diameter of 12 mm, which matches the microphone [24]. A wider cylindrical part begins at this level and has a fine internal thread at its lower end. The microphone can be screwed in if the protective cover is removed. The gold-coated diaphragm has a diameter of 12 mm and is fixed under the stable edge of the capsule. This edge has a height of 1.5 mm. When screwing the microphone into the juncture, the edge is pressed against the metal under the conical section, separated by a sealing ring. This junction is soundproof, but a tiny hole (not shown in the figure) allows pressure balance with the external air. With the use of special screws the capsule has to be carefully fixed to the microphone body in order to prevent any relative displacement. The cross-hatching in Fig. 1.21 shows the microphone. A stable cardboard tube separates the microphone body from the rigid foam, but the tube is stuck to the foam material. Thus, as there is "zero" space between the microphones and the cardboard, the tube provides good shock protection. Even when the halves of the dummy head are stuck together the microphones may be easily removed and replaced.

The "eardrums" are made of soft and porous foam material (grey area in Fig. 1.21). They are placed in the ear canal roughly at their natural position which is 28 mm from the canal entrance in the outer ear. The volume of the adjacent conical section between the eardrum and the diaphragm of the microphone is about 1.3 cm^3. Thus it is roughly equal to the volume of the human middle ear between the eardrum and the oval window at the entrance to the cochlea [29]. The eardrum of the dummy head shall, first and foremost, damp the Helmholtz resonances of the middle ear, that is, the conical section in the brass juncture. Second, it shall terminate the ear canal by an impedance which is at least approximately correct. Third, the eardrum shall be as sound-transparent as possible. This is essential for a good signal-to-noise ratio at high frequencies, including the required preamplifier.

In order to realize these goals, some cylindrical discs of various thicknesses were punched from porous foam material. This was done with punches of different diameters. The cylindrical model eardrums were pressed into the narrow section of the conical space in the brass juncture (see Fig. 1.21). They were then carefully stretched into shape in order to achieve a homogeneous distribution of the foam material. As mentioned above, the microphones could be easily removed and replaced. Thus, several foam cylinders could be readily compared with each other in order to optimize the eardrums.

The sound of the dummy head was preliminarily judged with head phones. Thus the punched cylinders could be sorted out immediately if the damping was insufficient. The next criterion was the audibility of a quartz alarm clock, which was placed on 4 mm of felt material lying on a table. The table was placed in another room, separated by a 24-cm brick wall and closed doors. With "good" eardrums and adequate amplification, the ticking of the clock was clearly audible with an impressively low noise level. Thus, the signal-to-noise ratio was regarded to be sufficient. These tests were carried out during a quiet night with no wind, in a house standing in a cul-de-sac road at the edge of a forest. Thus the best model eardrums combining the diameter, the thickness, and the porosity of the material could be selected with little effort.

The dummy head mount was then stuck into the neck. Its 6.35 mm thread fit a common microphone tripod, which supported the preamplifier, too. The preamplifier increased the signal level by 13 dB and took its power from the tape recorder. The first amplifying stage had to be carefully constructed. Thus the given signal-to-noise ratio of the microphones was nearly maintained, and transformer coupling was unnecessary. If, at an absolute sound pressure level of 124 dB, the microphones generated the maximum signal voltage, the preamplifier was not overdriven because it worked with ±12 V. As its output level was high enough, the signals could be transmitted asymmetrically through a 13-m cable. Even under extreme conditions no noticeable ghost signals were received by the overall system. (There were strong military radar beams from the Harz mountains before the "iron curtain" dropped in 1989.)

To employ the dummy head as a stereo microphone, we ensured a correct frequency characteristic of its preamplifier. However, first of all, we had to find the meaning of the word "correct".

Techniques for measuring sound pressure at the human eardrum—from a frontal plane wave source—are known from the literature [15, 17]. Unfortunately, for several reasons these data are not suitable for the adjustment of the preamplifier. First, the "eardrums" of the dummy head are not identical to the diaphragms of its microphones. Therefore, with regard to the frequency characteristic of the dummy head, the effect of the conical "middle ear" in the brass juncture is unknown. Second, data from the human inner ear regarding the movement of the oval window at the entrance to the cochlea are not available, and anyhow the diaphragm of the microphone might only roughly be compared with the oval window. Last but not least, the signals received by the dummy head will be retransformed into sound later on. Then the (new) sound waves have to pass another external ear and another ear canal before they are allowed to reach the eardrums of a person. Only then can the sound be processed in the auditory system! In short, data from the literature are no help when adjusting the preamplifier.

In a completely different approach it would be basically possible to optimize the frequency response of the preamplifier with the aid of listening experiments. However, this procedure is known to be difficult and time-consuming if exactness is desired.

Fortunately, subjective methods can be avoided by *self-related calibration* in the living room: The frequency characteristic of the dummy head can be adjusted using the dummy head itself. First, however, possible symmetry errors must be eliminated by a free-field measurement.

Symmetry errors become particularly obvious if the dummy head receives a frontal plane wave. Because no anechoic chamber was available for this measurement, the dummy head was placed in a garden when it was covered

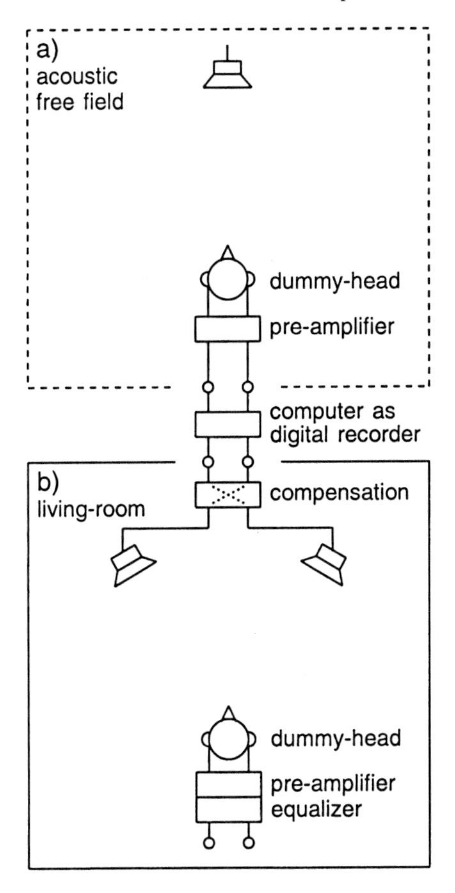

Fig. 1.22. Adjustment of the frequency characteristic of the dummy head

with snow. We used Lissajou figures to precisely adjust the dummy's nose to the loudspeaker, and then we compared the two "ears" (see Fig. 1.22a). After small electronic corrections, symmetry errors in the frequency range from 0.25 to 10 kHz stay below ±0.3 dB.

The measurement in the acoustical free field lead us directly to the self-related calibration and shall therefore be described in detail.

In the garden covered with snow the dummy head received a fairly plane sound wave from the front, generated by a directional loudspeaker box emitting sinusoidal warble tones at a distance of 3 m. The modulation range was 1/3 of an octave and the centre frequency was increased step by step from 50 Hz to 20 kHz. A controlling microphone was mounted in the front at a distance of 20 cm from the face. Thus the same sound pressure could be adjusted at each centre frequency of the warble tones. The small deviations in the frequency characteristic of the microphone were taken into account and eliminated from the start [26]. The signals received by the dummy head were digitized and stored in the computer [14] which includes a digital-to-analog converter for the playback as well. The analog circuits in the computer had to be revised in order to guarantee error-free operation.

The recording from the garden was then reproduced in the living room. The same dummy head was placed at the listener's location, and the compensation filter was switched on (Fig. 1.22b). At each centre frequency the preamplifier should produce quite the same signal level as in the garden. However, this ideal case does not occur automatically. In order to remove the differences, a commercial equalizer was employed. Its 2×33 band-pass filters cover the full audio range and have a bandwidth of $1/3$ octave each [30]. When the level adjustments were regarded as final all sliding controls of the equalizer had to be replaced by fixed resistors, thus excluding any accidental readjustment.

The transmission in the living room may be described by the relation $S(f) \rightarrow S(f) \cdot T(f)$, where f denotes the frequency; phases are disregarded as they are unimportant in this case. The spectrum $S(f)$ of the stored computer signal is multiplied by the transmission function $T(f)$ which includes first the special reproduction with compensation and second the effect of the dummy head with the preamplifier and the equalizer. The amplitude part of the transmission function $T(f)$ is plotted in Fig. 1.23. The curve may be calculated as the level difference between the signal $S(f) \cdot T(f)$ at the output and the signal $S(f)$ at the input of the living room. The deviations in the frequency characteristics of the loudspeakers are eliminated arithmetically. In essence, the purpose of adjusting the equalizer is to arrive at the straight 0 dB-line in this diagram.

The meandering line in Fig. 1.23 roughly approaches the desired result. Thus the transmission factor is $T(f) \approx 1$, except for a dropping tendency at low frequencies. However, this roll-down was constructed purposely because

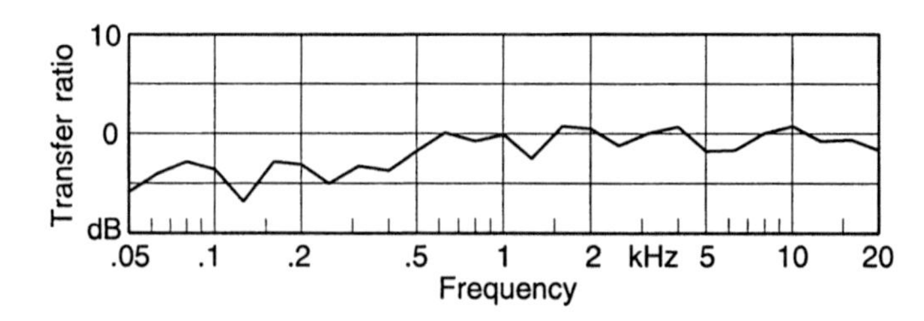

Fig. 1.23. Dummy head in the living room receives signals from the garden, compensation filter switched on

the dummy head will be used for stereophonic recordings in closed rooms, rather than as a measuring device. Consequently an accentuation in the bass range has to be taken into account. The proximity effect known from directional microphones has quite the same tendency, and special counteracting bass controls are installed in some microphones. Thus the roll-down at low frequencies need not be regarded as a disadvantage. It all depends on the purpose for which the dummy head will be used. The bass controls known from common stereo devices are well suited for adjusting the final frequency response as required.

The irregular peaks and dips of the curve indicate methodical inaccuracies: the measured values were combined in the form A–B–C which is prone to errors. These errors can be avoided. Plotting the level difference of the signal $(S \cdot T)$ and the signal S means calculating the logarithm of the quotient $(S \cdot T)/S$. As the fraction may be cancelled down, of course, the result does not change if another input signal S is used. We can thus realize that Fig. 1.23 describes *only the transmission in the living room*, but not the recording from the garden.

This important point—that the general tendency of the curve does not depend on the signal—can be demonstrated as well as exploited experimentally. We can restrict the discussion to the lower part of Fig. 1.22 and present 1/3 octave band noises from a test CD to the dummy head in the living room [13]. The essential transmission curve can thus be checked, possible errors can be detected, and the final adjustment of the equalizer can be revised. The boldface curve in Fig. 1.24 shows the signal levels measured at the dummy head, again including the preamplifier and the equalizer as mentioned earlier. At all centre frequencies of the 1/3 octave band noises, the test CD supplies the same signal level (except for a slight decrease at high frequencies which is eliminated). The compensation filter was switched on again. The boldface curve deviates from the straight 0 dB-line just as the dotted line, which was

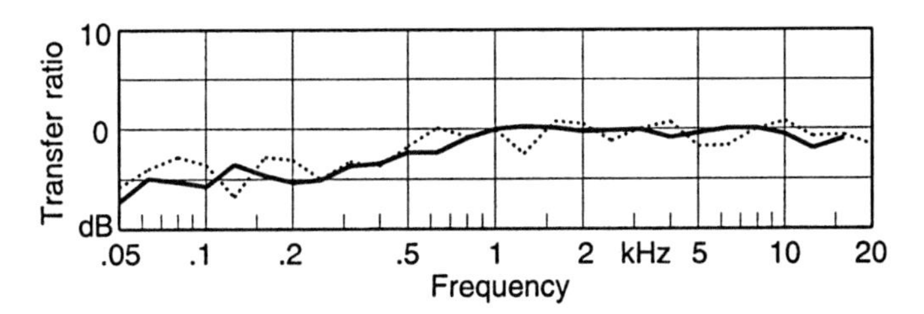

Fig. 1.24. —— Dummy head in living room receives 1/3 octave band noises, constant input level at the stereo setup, compensation on. · · · · See Fig. 1.23, recording from the garden

taken from Fig. 1.23 for comparison. Once again the faults in the frequency response of the loudspeakers are eliminated. Thus all data for the boldface curve could be measured right away in the course of one hour. This reduces error. As expected, the two curves agree in the main. They are both valid for one and the same adjustment of the equalizer.

1.7 Conclusions

Except for the roll-down at low frequencies, the reproduction with the equalized dummy head in the living room (Fig. 1.22b) has the transmission factor $T \approx 1$. This is discernible in Fig. 1.24; *and if the recording, too, is made with the equalized dummy head it is still valid.* The reader may imagine the equalizer already inserted in Fig. 1.22a. What are the logical consequences of this special case, that the equalized dummy head in the living room "listens to its own recordings"? The signal levels of the recording are reproduced at the equalizer output in Fig. 1.22b, except for the deviations from $T = 1$. Therefore, because we have the same dummy head in the living room as in the recording process, the input signals of the preamplifier are also equal. So are the sound pressure values at the eardrums. Therefore, the *overall transmission* is perfect for the dummy head listening to its own recordings, and this is the result of the self-related calibration.

The last but decisive conclusion requires the presumption that the human head equals a dummy head, but only geometrically and only up to the eardrums (let us hope so!). If this equivalence is given, the sound waves in the living room do not change if the listening dummy head is replaced by a human head. This consideration is valid in general. The dummy head may be exposed to any simple or complex sound whatever in the recording process: The signals at its eardrums will be reproduced at the eardrums of the person in the living room.

The equalizer adjustments described earlier are not only sufficient but also necessary for the perfect transmission. This can be shown by a simple gedanken experiment: If, at a certain frequency, the amplification of the equalized preamplifier is disadjusted by +3 dB, for example, the recording level at that frequency will be increased by 3 dB. When the same dummy head "listens" to the recording, the level at that frequency will be increased once more by 3 dB, reaching +6 dB. Thus the two levels of the recording and the reproduction are no longer equal, and the new adjustment is incorrect.

With the use of a test CD the overall frequency characteristics of the dummy head can be adjusted rapidly and precisely while the compensation is switched on. The idea that this adjustment might be irrelevant for the recording process, only because it was achieved in the living room, proves to be false. Quite to the contrary, the transmission must be regarded on the whole. Perfect *reproduction* is precisely what is necessary and sufficient for the perfection of the whole system.

This brings us to a summary of the whole first chapter. Thinking of the recording process and the reproduction as separable should be avoided in general. The signal from an excellent condensor microphone, for example, is well suited to monophonic reproduction. However, if this basic microphone-to-loudspeaker system is merely duplicated, one immediately pays dearly for the known delightful stereo effect with undesirable side effects. The acoustical interference at the head of the listening person spoils the sound spectrum and destroys a large part of the spatial information. The defect can be removed by additional compensating signals which account for the geometry of the head. When reproducing dummy head signals this error compensation is required, too.

The strangely twisted shape of the external ear is a biological fact which must be accepted. This peculiar thing at the head resembles a keyhole in an old door—it encodes and decodes the stereophonic information. Only signals that possess the proper key can open the wide subjective sound space.

Part II

The Hearing Process in Concert Halls

2

Powerful Onset of Reverberation

2.1 Introduction and Definitions of Terms

When investigating the acoustics of a concert hall it is common practise to excite the room by a shot from the stage. A microphone placed in the room receives the "impulse response", which is usually stored for later evaluations. Its oscillogram shows a large number of separate pulses that grow in density over time. To explain these pulses, we may consider sound propagation in the form of rays. The first pulse is caused by the "direct sound" which travels along a straight line from the stage to the microphone. All the later pulses are called "reflections", as the sound path includes a detour with at least one reflection. The actual reflections of the sonic rays take place at the walls or the ceiling. The audience may be imagined to be sound-absorbing.

The "onset of reverberation" is caused by the direct sound and a few early reflections. The genuine reverberation consists of the large multitude of later reflections, most of which have travelled over long zigzag paths. Because of the reflections' large number and high density in time, the reverberation appears to decay smoothly. The limit between the onset of reverberation and the genuine reverberation is variable, and it depends on the style of music presented. However, these details are unimportant in this chapter. We may assume that the first 50 ms of the impulse response make up the onset of reverberation, immediately followed by the genuine reverberation.

The duration and also the sound colour of the genuine reverberation are important quality factors in any concert hall, and there are many scientific publications in this field. An obvious and important property of the genuine reverberation is its smoothing effect, comparable to a piano pedal. However, the experiments described in the following concern the *onset of reverberation*. This topic is fascinating because the early reflections can open up strikingly wide sound impressions in the front while still maintaining high definition. Good halls can be distinguished from bad halls just by these details. Therefore we shall investigate how these subjective effects are generated by a few specific early reflections and their interaction. In this field of acoustic research only some tentative approaches are known to the author.

One of the first approaches was meant to simplify the task. It is based on the fact that the analyzing power of the ear is limited. If, for example, *only one* reflection is eliminated from the great multitude forming the reverberation, the ear might be unable to notice the small difference. The same idea was applied to the onset of reverberation, and the Absolute Threshold Of Detectability was measured with early reflections (in German this is called **a**bsolute **W**ahrnehmbarkeits **s**chwelle, or aWs: [1]). For this purpose, simple sound fields with a direct sound and only two reflections are simulated in an anechoic chamber. A subject has to analyze a given sound, continuous speech, for example. In relation to the direct sound from the front, the first reflection has a fixed direction, a fixed delay, and a fixed sound pressure level. The second reflection arrives from another direction. At first it has an extremely low volume and thus it is too feeble to be perceived. While the reflection is alternately switched on and off, its sound pressure level is increased until a difference can be noticed at the moment of switching. In the opposite case, the sound pressure level is reduced until the difference has vanished. This process is carried out until the threshold of detectability has been determined with the desired accuracy. The measured threshold value depends on the direction as well as on the delay time of the reflection.

When investigating a real room, these experimental results can help to exclude some of its reflections from the further discussion because they have no detectable effect.

Another extreme case—in contrast to the inaudibility of reflections—was investigated by H. Haas in 1951. He increased the sound pressure level of a single simulated reflection until its loudness appeared to be equal to that of the direct sound. This condition is reached if the sound pressure level of the reflection is 10.4 dB higher than that of the direct sound [2]. The result is rather independent of the delay time and direction of the reflection.

In the impulse responses of real rooms there are no such powerful reflections. Nevertheless, the effect can be used to support the voice of a speaker by a reflection which is simulated with a loudspeaker. At a relative sound pressure level of +6 dB and a delay of 10 ms, for example, the reflection is not experienced as disturbing.

With these questions as a starting point, the subjective effects of a few early reflections should be investigated quite generally. In the following, we discuss the experimental setup and the measuring procedure.

2.2 Sound Fields According to Haas

In a large anechoic chamber the reader may imagine 65 loudspeakers on a hemispheric surface with a radius of 2.70 m, the surface being domed to the top. The loudspeakers are evenly distributed on the surface and point to the centre of the hemisphere. Figure 2.1 shows a photograph of this arrangement; its schematic is shown in Fig. 2.2, but with equi-areal projection to the horizontal plane. Each point marked in this plan indicates a loudspeaker and also a direction in space. The subject is seated in an armchair, and thus the head is placed in the centre with sufficient accuracy. The photograph and the plan show that the loudspeakers are arranged in five horizontal circles. Twenty loudspeakers are placed on the lowest circle. The small elevation angle of this circle results from the subdivision of the hemispheric surface and does not interfere with the experiments. The head symbol at the centre indicates the place for the subject and the line of view, which defines the azimuthal

Fig. 2.1. Anechoic chamber with hemispheric arrangement of 65 loudspeakers

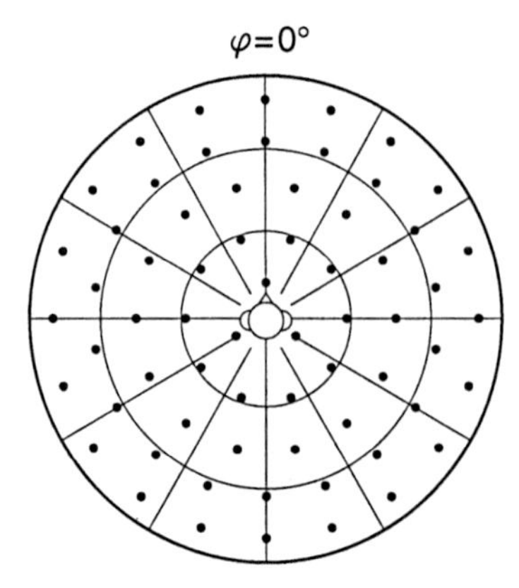

Fig. 2.2. Hemispheric dome of 65 loudspeakers in an anechoic chamber, with the place of the subject in the centre; equi-areal projection

angle $\varphi = 0°$. The subject is handed the plan upon entering the anechoic chamber. He or she acquaints themselves with the plan's meaning and takes a seat. From this point forward the head should not move. However, the geometrical impression is supported by about 20 loudspeakers that permanently remain in the field of vision.

First, we scrutinized the effect of a single reflection in a series of three simple sound fields. The reflections were generated with the use of one or two loudspeakers in the lowest row. Each sound field was judged by a group of subjects working one by one, listening to continuous anechoic speech from a tape. The first "sound field" was extremely simple. Only one loudspeaker was used and generated a direct sound at the azimuthal angle $\varphi = 0°$. In the next two sound fields a reflection was added, either at the azimuthal angle $\varphi = 36°$ or at $\varphi = 108°$. For increasing the overall accuracy, the mirror-inverted arrangements were also investigated, and the results were combined.

The subjects had to mark all the directions in which they could perceive the speech. Their working time was unlimited. The next sound field was presented only after request by the subject. Figure 2.3 depicts the results of 21/15/21 subjects. The arrows in the diagrams indicate the selected loudspeakers in the lowest row. The delay times and the sound pressure levels are related to the direct sound.

For reference, Fig. 2.3.1 describes the fundamental case of just one loudspeaker working at $\varphi = 0°$ in front of the subject. This direct sound had an absolute sound pressure level of 70 dB, measured at the subject's seat. All the levels mentioned in the following are related to this value. In the other two cases a simulated reflection was added. It was generated by a second

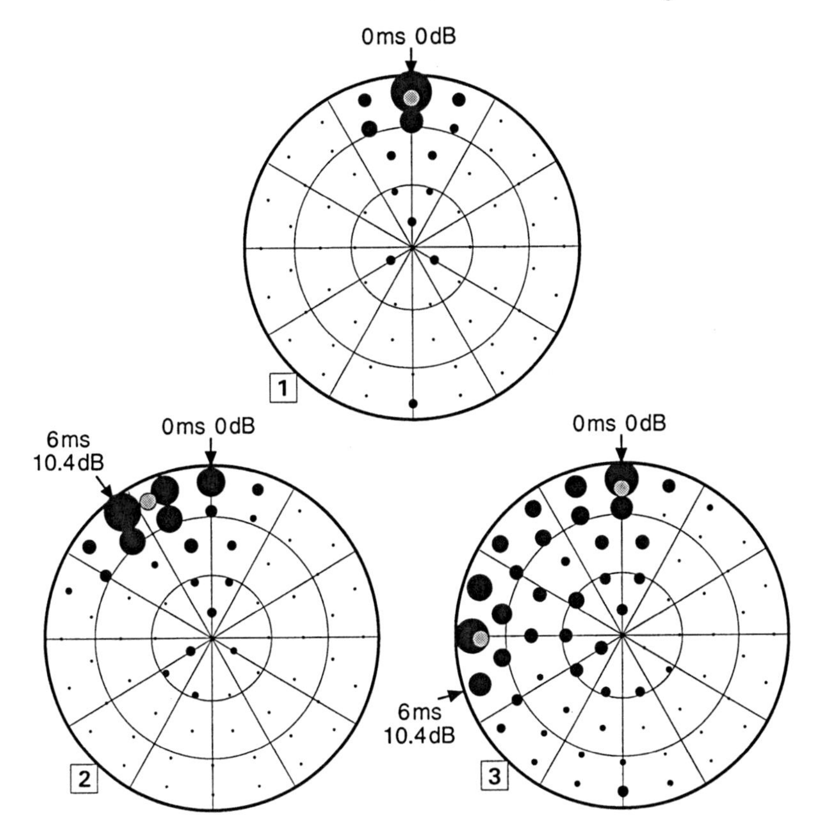

Fig. 2.3. Sound images concerning the Haas effect. Marked talker's position: ◍

loudspeaker in the lowest row; the same speech was sent, but was delayed by 6 ms, again in relation to the direct sound. This simulated reflection had a relative sound pressure level of +10.4 dB, which is the value measured in the famous Haas experiment [2]. The sound directions perceived by the subjects are indicated by black circles. Their area is proportional to the percentage of the subjects who marked that direction. If the percentage is zero, then the direction is shown by a tiny dot. Figure 2.3.1 contains about two perceived directions per subject. In the third experiment, each subject marked about six directions which are concentrated near two separate centres. Figure 2.3.2 shows an intermediate case. Its imbalance might be explained by the limited accuracy of ±2 dB for the value +10.4 dB measured by Haas, or by imprecise answers from the 15 subjects.

Both Figs. 2.3.2 and 2.3.3 concern the Haas effect, but the delayed sound component arrives from quite different directions in the horizontal plane. In both cases the delayed sound is very powerful, and yet the directions subjectively perceived can be assigned to the two objective sound components in equal parts. The following additional task should make the subjects decide clearly: "Please mark the place of the talker whose voice you hear!" Of course, the answers are again statistically distributed. However, the grey spots in the figures are meant to indicate only the centres of these distributions. The direction marked by the spot either falls between the two physical sound directions [3], or the perceived voice is statistically split.

The brief series of experiments illustrates the aforementioned experiments from the years 1951–1952. At that time it was asked whether the two sound components were perceived as equally loud, or whether the voice was perceived right in the middle between the two physical sound components. The subjective sound distributions were not yet measured at that time. The experiments took place on a flat roof in the open air, and the subject could see only two loudspeakers. Thus it was obvious which sound components had to be compared.

In a concert hall the situation is much more complicated, although the concertgoer can see the musical instruments placed on the stage. Thus the directions of all the *direct sound* components are quite obvious, but the listener cannot detect the physical sound *reflections* determining his subjective impression. The extreme case of a noticeable "echo" means "bad acoustics" and may be excluded right away; and the special situation of experienced acousticians or musicians regarding the walls in order to detect sound reflecting surfaces can also be disregarded. However, if the reflections are not separately perceivable their subjective effects cannot be mutually weighed. In the search for a starting point in this intricate situation we might take another look at Figs. 2.3.2 and 2.3.3: The centres of the two distributions—as well as the grey spots—are not unambiguously situated in the 0°-direction. This is a succinct hint at a well-known rule: discrepancies between optical and acoustical impressions are confusing and should be avoided.

2.3 Drift Thresholds and Subjective Sound Distributions

For a listener in a closed room the acoustical impression definitely depends on the reflected sound components, that is, on "the reflections". Their sequence and their directions determine "the acoustics" of a concert hall, a theatre, or any other room. However, the auditory process that creates a spacious impression from the reflections is nearly uninvestigated. Even for a single sound source in the acoustical free field, our knowledge of the directional perception could only recently be extended to the full 360° range [18]. Much less is known about the way in which the auditory system processes the stream of data in a complex sound field. Therefore, a physicist would still consider the effects of the reflections in a generalized way. This may be indicated by a few key words and phrases: law of the first wave front [31]; early energy ratio [4, 5]; loudness function [6, 7]; and interaural cross-correlation [28, 32, 33].

In order to consider the situation more closely, we explored step by step the emergence of a subjective sound image from a direct sound and a few early reflections. The experimental series began with the direct sound and just one reflection and ended with six reflections. Thus the *onset of reverberation* was investigated, while the genuine reverberation was excluded. The experiments were again carried out in the anechoic chamber with loudspeakers in the lowest row of the hemispheric arrangement. The subject received the direct sound from the front (azimuthal angle $\varphi = 0°$). The absolute sound pressure level of the direct sound was 68 dB, measured at the subject's seat. This value serves as a reference in the following. Once again continuous speech, recorded with no reverberation, was used for the measurements. Continuous speech is well suited for this experiment because of its impulse content. Music is unsuitable because its various styles, the way of playing, the different instruments, etc. require undesirable distinctions.

The first sound field consisted of the direct sound and a single reflection at the azimuthal angle $\varphi = +36°$. Its delay could be chosen in the range of 6–108 ms in relation to the direct sound. Having selected the delay time the subject took a seat in the anechoic chamber. With a sliding control fixed at the seat, the level of the reflection could be varied.

The following question had to be observed during the adjustment: Does the direction of the greatest loudness agree with the line of view, and is there no echo to be noticed? If the answer was "yes" the level had to be increased; otherwise it had to be decreased. In the course of the adjustment the question had to be observed more and more carefully, and the level had to be varied less and less until the question could be answered neither by "yes" nor by "no". Then the subject (the author) had to leave the anechoic chamber in order to read and to note down the adjusted level value. There was no possibility of reading the level value during the adjustment process.

Figure 2.4 plots the measured levels valid for the 36°-reflection as a function of the delay. The curve may be termed a "drift threshold" (DT). Its first point is for a delay time of 6 ms. At this delay value the highest permissible level of the reflection is +1.1 dB; otherwise the perceived centre of the voice would drift sideways, and thus leave the line of view. In order to comment on the word *drift* we may recall the situation in a concert hall: the musical instruments should be heard where they are seen.

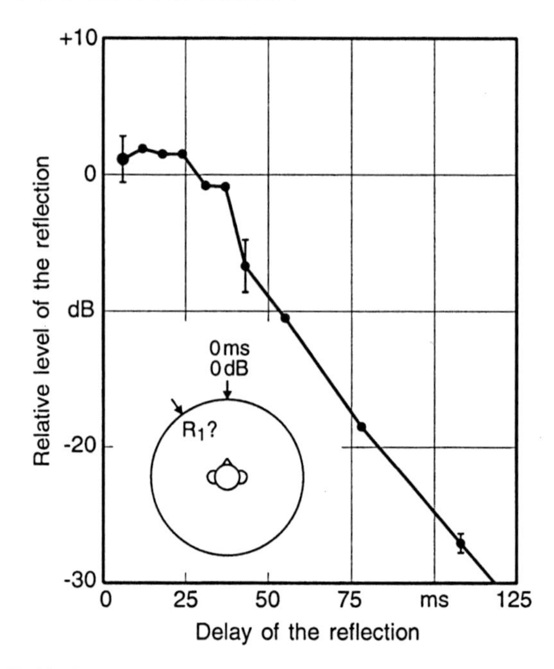

Fig. 2.4. Drift threshold (DT) of a first reflection R_1, continuous speech

When the delay time is increased the drift threshold remains nearly constant in its first part and then drops off steeply. With regard to the reflections in concert halls, level values above 0 dB usually need not be expected if the sound is not focused by concave surfaces. However, early reflections at a level just below 0 dB—that is, near the drift threshold—are realistic and very desirable.

The general shape of the curve resembles the Absolute Threshold Of Detectability, measured in 1961 with a similar experimental setup [1]. However, the drift threshold is higher by 20–30 dB. Any reflection at the DT-level will therefore clearly affect the sound image.

The basic subjective effect of such a reflection should of course be explored. Therefore the delay time was fixed at 6 ms, which means a detour of about 2 m in the context of sound in a concert hall. The relative level of the reflection was adjusted to the drift threshold, which is at +1.1 dB. The sound field thus specified was judged by a group of 21 subjects working one by one. Once again they had to mark the perceived sound directions in the circular plan. The result is shown in Fig. 2.5. Obviously the 36° reflection at the drift threshold causes a widening of the sound image (compare with Fig. 2.3.1), and yet the perceived centre of the voice remains on the line of view (grey spot).

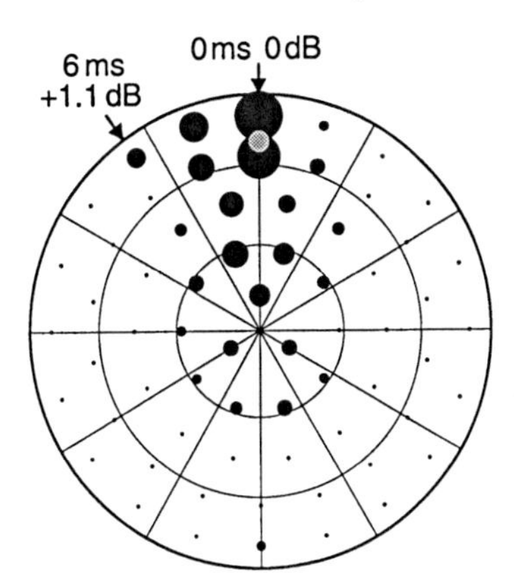

Fig. 2.5. Subjective sound distribution and perceived centre ● of the voice, measured with a reflection R_1 at its DT-beginning (21 subjects)

This point, indicating the perceived talker's position, is placed neither at the circular edge of the image nor on the largest ring of loudspeakers. Its distance from the circular edge represents an elevation angle which may be interpreted as follows: The subjects began their work by marking a certain group of points. In a search for the centre of the voice in this ensemble they looked for its "centre of gravity" which, logically, cannot be found at the edge.

Judging the elevation angle of the perceived voice, we might consider the simplest case of Fig. 2.3.1 again: Even with a needlelike sound ray from the front, the image extends upward, and the centre of the voice is perceived as "too high". A possible explanation for this known phenomenon is the fact that, with symmetry of the head presumed, a sound ray from the front can cause no interaural signal difference whatever, and especially no interaural time delay. Thus, when determining directions, the aural evaluation process is less precise, as it works monaurally.

The described effect may be regarded as unimportant and is no longer discussed here. Thus the term *drift* threshold remains restricted to the horizontal aspect which corresponds to the basic meaning of the word originating from sailors' slang.

For increasing the accuracy and reliability of Fig. 2.5, the measuring procedure was repeated with a mirror-inverted reflection at $\varphi = -36°$ (half right). The two results, valid for $\pm 36°$, were combined for the diagram, but they are shown as if only the $+36°$ angle had actually been used. This procedure was followed quite generally: *The subjects had to judge each sound field in its mirror-inverted version as well*, and the results were combined in the evaluation.

In most cases the measurement of the drift thresholds was repeated in the same way, but only the author himself was available for this time-consuming and stressful work. For achieving a sufficient overall accuracy just the same, the threshold curve had to be remeasured several times in the course of a few days. For example, the important first point of the curve in Fig. 2.4, at a delay time of 6 ms, is an average of 24 results which were acquired in the course of three days. In this case the threshold level is +1.1 dB, with a standard deviation of ±1.7 dB. The result is supported by the sound image shown in Fig. 2.5. The perceived centre of the voice has not drifted—it is located at $\varphi = 0°$. Thus a group of 21 subjects confirms the drift threshold which was carefully measured by one person.

The second experiment of the series is described in Fig. 2.6. Now the first reflection R_1 is fixed at 6 ms/+1.1 dB, which is the first point of the first drift threshold (thin line). For the time being, the variable reflection R_2 was added at $\varphi = -36°$ on the opposite side of the head. Using the procedure outlined above, its drift threshold was measured in the delay range 12–108 ms (upper curve).

Drift thresholds would be less important if they depended on directions. For determining whether this is the case, the curve was remeasured for some

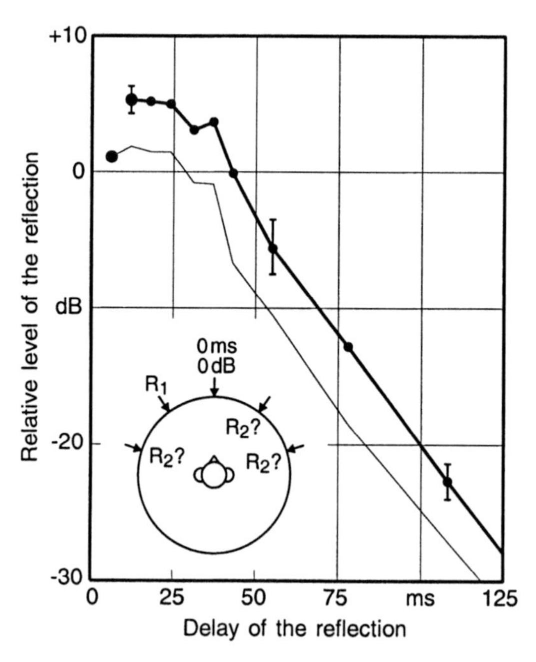

Fig. 2.6. Drift threshold (DT) of a second reflection R_2, continuous speech

other directions of the reflection R_2. However, *no directional dependence could be detected(!)* within the measurement accuracy. In Fig. 2.6 the directions used for R_2 are indicated by question marks near the head symbol ($-36°/+72°/-72°$).

The measured drift threshold (upper curve) results from six partial measurements. Its first point—at a delay of 12 ms—shows that the highest permissible level of the second reflection is +5.3 dB if the perceived centre of the voice is to remain at $\varphi = 0°$. The curve resembles that of the first reflection (thin line) but is shifted toward higher levels.

The subjective sound image was explored with 20 subjects. It could be determined only with the arrangement and the set of data specified in Fig. 2.7, but including the mirror-inversion. Once again, this case forms the basis for the following experiments of the series.

The sound distribution covers a rather large solid angle. The perceived centre of the voice shows virtually no drift, although the second reflection has a higher level than the first one. It is obvious and yet surprising that the strong reflection from the right "pushes" the sound image a little to the left. However, the first reflection must have a stabilizing effect, as the overall image remains rather symmetrical. The pushing phenomenon is probably not a measuring error, as similar effects are known from the literature.

The subjects are quite sure to perceive sound in the range above their head, and the same effect appeared already in the first experiment of the series (Fig. 2.5). However, all loudspeakers in that region of the anechoic chamber are mute, and the intensity of sound portions reflected by their membranes can be regarded as zero. The sound perceived above the head is thus of purely

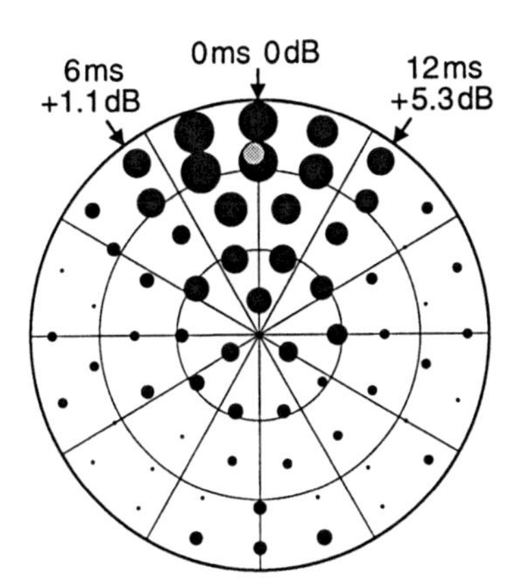

Fig. 2.7. Subjective sound distribution and perceived centre ◉ of the voice, measured with two reflections R_1, R_2 at their DT-beginnings (20 subjects)

psychological nature, created as a side effect of the data processing in the listener's head.

The reflection R_1 at a level of +1.1 dB maintained a certain relation to reality. This relation is less distinct with R_2, as levels near +5 dB do not typically occur in the impulse response of a concert hall.

The third experiment of the series is described in Fig. 2.8. As mentioned earlier, it is based on the second experiment. Now there are two reflections which are fixed at the first point of their respective drift threshold. R_1 is fixed at 6 ms/+1.1 dB, and R_2 is fixed at 12 ms/+5.3 dB. The newly added reflection R_3 is variable. As its drift threshold might depend on the directional arrangement of the reflections, two quite different cases were tested. In Fig. 2.8 they are shown with circular sketches below the curves. At the edge of the left circle a zigzag arrangement indicates the reflections R_1–R_3, whereas the right circle shows R_1 and R_2 on one side of the head, and only R_3 on the opposite side. However, even with these extremely different cases, *no dependence on the directions could be detected within the measurement accuracy (!)*. All four partial results are contained in the upper curve, including

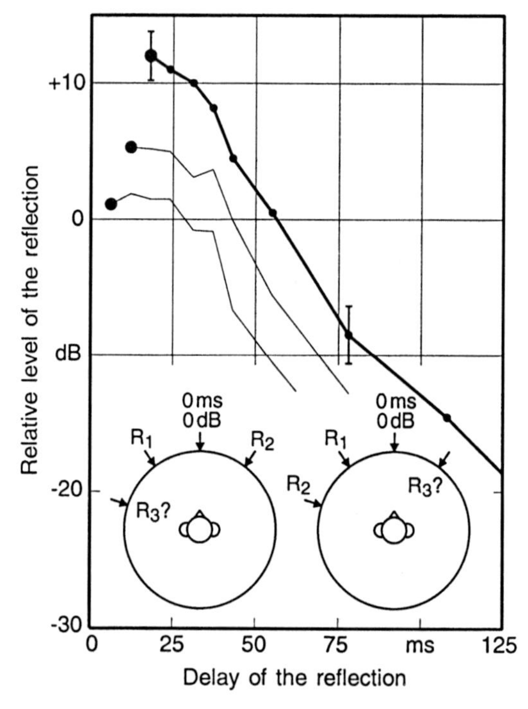

Fig. 2.8. Drift threshold (DT) of a third reflection R_3, continuous speech

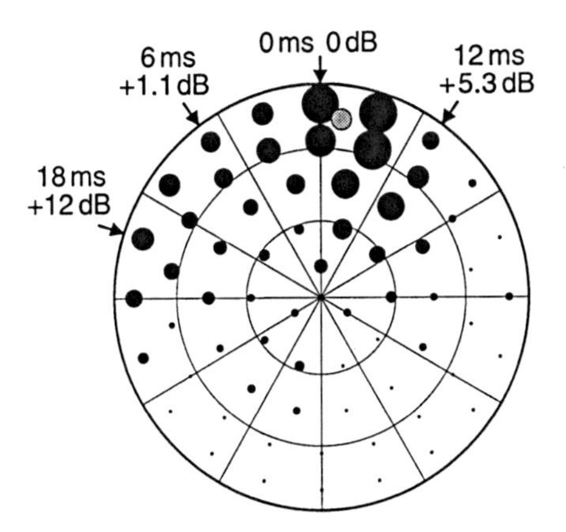

Fig. 2.9. Subjective sound distribution and perceived centre ● of the voice, measured with three reflections R_1–R_3 at their DT-beginnings (22 subjects)

the mirrored arrangements. The discussion of directional arrangements will be continued in Sect. 2.5.

In this case (Fig. 2.9) the subjective sound distribution is not as clear as with two reflections. The sound image is wider, but the 22 subjects do not quite agree on that aspect. The lateral shift of the perceived talker's position is not significant. However, the third reflection is stronger than in the Haas effect, in which a drift is expressly *accepted* (see Fig. 2.3)!

No statistical splitting of the perceived voice appeared in this investigation (compare with the Haas effect, Fig. 2.3.3). Once again, however, the perceived centre of the voice (grey spot) is pushed away from the last strong reflection. Will the left part of the image be refilled by the next reflection? Will it again be possible to achieve a better equilibrium? Will the next thresholds, measured by one person, be falsified by the groups of subjects?

As further reflections R_4–R_6 were added, the drift thresholds became unexpectedly high. This overloaded the amplifiers. Unfortunately it was impossible to repeat the whole experimental series at a reduced loudness right from the start, because unnecessary stress should be avoided when working with groups of subjects. Therefore all signal levels had to be reduced by 6 dB before the fourth reflection could be added. Further reductions by 6 dB were necessary for the fifth and sixth reflections. These changes are important for the theoretical evaluation (see Chap. 4). After the fourth reflection was added the thresholds were no longer checked for independence of directions.

Figure 2.10 shows that the series of experiments ended at six reflections, which zigzag as indicated below the pile of curves. All curves are clearly staggered one over the other and have rather similar shapes. Their

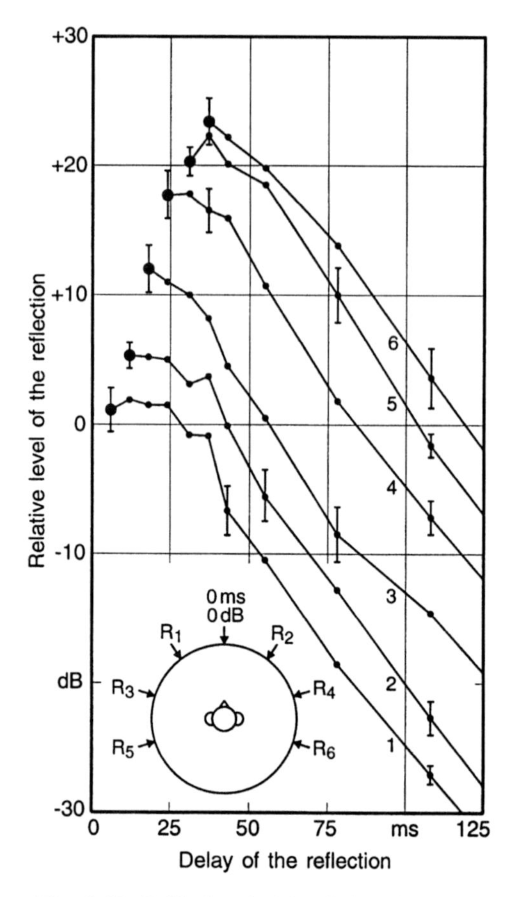

Fig. 2.10. Drift thresholds of six reflections

first points show a nicely rising succession. These points are used when fixing the data of the reflections in order to measure the next drift threshold.

The first point of the highest curve indicates that the level of the sixth reflection may exceed the level of the direct sound by 23.4 dB, and yet the talker will be perceived in the 0° direction! As mentioned earlier, the directions of the reflections seem to be unimportant as long as they are different. This was shown with up to three reflections. Two special series of measurements for the Haas effect revealed a directional dependence of ±1.3 dB or ±0 dB, according to the chosen arrangement.

In Fig. 2.11 the sound distributions are presented in their context. Each of the diagrams shows the plan of the hemispheric dome of loudspeakers in the anechoic chamber, and each diagram contains the listening results of

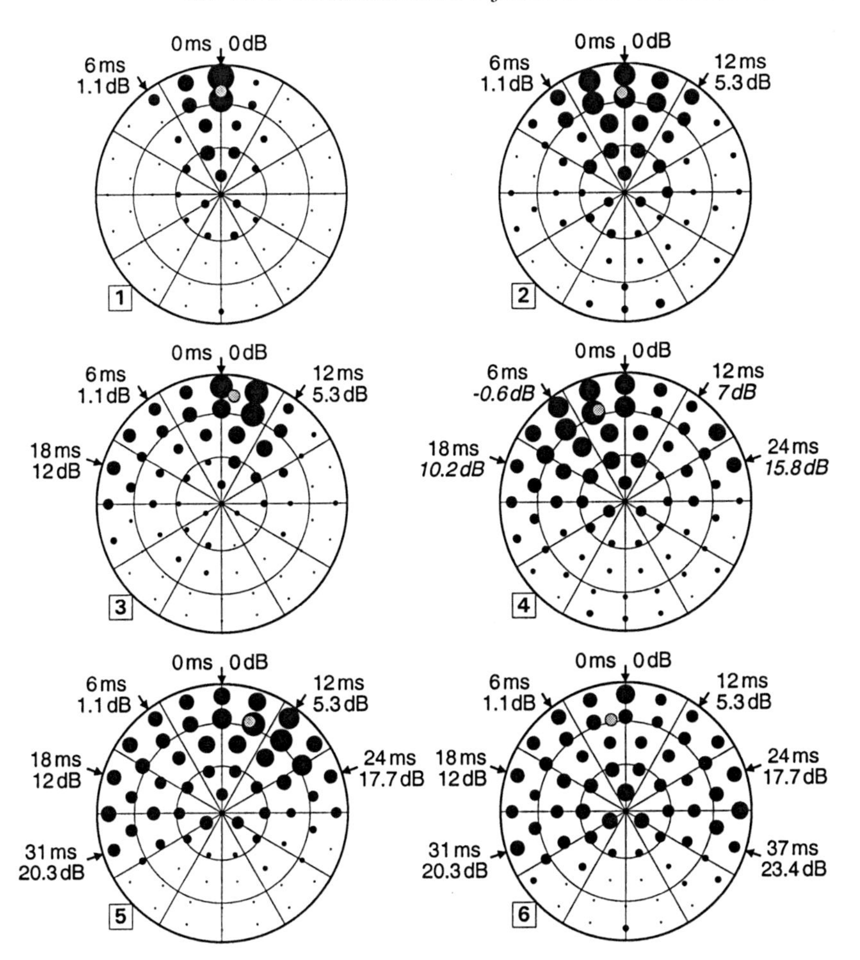

Fig. 2.11. Subjective sound distributions with anechoic speech, all reflection levels at drift thresholds (except diagram 4). Perceived centre of the voice: ◍

a group of 21–28 subjects. If a subject perceived sound from a certain direction, he or she marked it in the plan. The area of the small black circles is proportional to the percentage of the subjects who marked that direction.

The first little diagram was determined with the direct sound and just one reflection. The series ends with six reflections. Their zigzagging sequence is intended to remind us of reflections in a concert hall. The delay times and levels are adjusted in accordance with the beginnings of the drift thresholds, but the levels in Fig. 2.11.4 are an exception as they are lower by about 2 dB.

2.4 Discussion of the Experimental Results

Initially the groups of subjects were meant mainly to check whether the drift thresholds were measured correctly: $\varphi \approx 0°$ for the grey spots. However, the series of diagrams show an ever-growing diffuseness. This pleasant side effect was not quite unexpected. The last partial diagram, for example, contains 55 perceived directions; and because of the precise questioning, the directions actually marked are only a subset in an infinite set of perceived directions. The subjective sound distributions extend over wide ranges of the upper hemisphere, and perceived directions right above the head seem to be quite normal. To avoid confusion we must remind the reader of the fact that only seven loudspeakers in the lowest row were active. This again proves that the perceived (subjective) directions do not have to agree with the physical (objective) sound directions.

The following sentences concerning diffuseness are contained in a textbook on acoustics [34]: In a concert hall the sound shall reach the listener from many directions, so to speak in a surrounding manner, but not as strongly as from the front. In halls with smooth walls and a flat ceiling this demand can hardly be met. Therefore the enclosing surfaces should be "roughened up", thus achieving more or less diffuse scatter of a simple sound wave in a solid angle as large as possible, but avoiding reflection according to the reflection law.

We may notice from this quotation that the physical sound directions and the subjectively perceived directions are regarded as equal without thinking. However, the experiments described here show that this is inadmissible. Our experiments show that the demand to "roughen up" the enclosing surfaces might be unnecessary. In principle the diffuseness demanded in the quoted sentences may well be achieved by a couple of unshattered (strong) early reflections.

Figure 2.11 leads to the conjecture that maximum diffuseness is achievable only if all the levels in a series R_n of reflections are adjusted to their respective drift threshold. In this case the left-right equilibrium is always maintained, while larger and larger amounts of sound energy, arriving from more and more directions, are "trying" to disturb that equilibrium, thereby increasing the diffuseness.

If, at this point, the level of a reflection R_N is reduced (case A) the drift threshold of the subsequent reflection R_{N+1} will automatically be lower. As a consequence, to avoid a drift of the perceived centre of the voice, the level of the reflection R_{N+1} has to be reduced too, and so on. Thus the diffuseness decreases. However, if the reflection R_N is too strong (case B) the perceived centre of the voice would drift anyway. Thus the goal of maximum diffuseness would likewise be missed.

The final sound field with six reflections is particularly impressive. The voice appears outside the head and surrounds it nearly completely. Yet it seems "dry", like a voice under the bedclothes. (This strange sound might be meant in the famous science fiction novel "1984" written by G. Orwell: "big brother is watching you".) Although the sound impression is very diffuse it shows a *lack of depth*. Thus there is no spacious impression in the meaning of the word. The effect might be provisionally explained by the idea that the ear is unable to relate the "reverberation time" 37 ms with spatial dimensions. In any case it can be concluded from this experience that the terms diffuseness and spaciousness do not mean the same thing. A definition of the term diffuseness will be given in the next chapter.

The subjects had the additional task of marking the perceived centre of the voice. As expected the answers are once again statistically distributed. The grey spots in the diagrams stand for the centres of these distributions. In this task the subjects did not simply search for the "centre of gravity". If they had done so the grey spot would have moved upwards when the sound extended to larger and larger regions, including the range right above the head. However, the centre of the voice remains at an elevation angle of 15–30° (see Fig. 2.11). The perceived azimuthal angle of the voice is obviously fixed by an axis which is established by equal sound weights on its two sides. In two partial diagrams this axis deviates by more than 10° from the line of view. However, the perceived centre of the voice did not *in general* shift sideways. Thus the measured drift thresholds prove to be at least roughly correct.

The experimental series described in this chapter differs from the Haas experiment in the criterion applied: *zero drift* $\leftrightarrow$ *equal loudness*. As found out by Haas, a single reflection, delayed by 6–23 ms, is perceived to be just as loud as the direct sound if its relative sound pressure level is +10.4 dB. In this case the centre of the voice drifts sideways, or it comes to a statistical splitting (Fig. 2.3). In Fig. 2.11 the situation is quite different. The fact that the level values form an increasing series causes no relevant drift of the perceived centre of the voice. The centre remains on the line of view although the relative level of the last reflection is +23.4 dB (see Fig. 2.11.6). The sound is perceived as omnidirectional, and all components are perceived as equally loud.

This should be explained in more detail: When marking the directions of the perceived sound, the subjects were not requested to weigh the loudnesses of different components. They simply had to make various yes–no decisions, but the evaluation showed that even this simple task caused problems for many subjects. Some of the 28 subjects working at Fig. 2.11.6, for example, marked only three or four of the 55 directions documented for the group. However, it may be assumed that the sizes of the black circles represent the loudnesses of the marked components. Equal loudnesses serve as a basis for Fig. 2.3, too. Thus, when comparing the two experimental procedures, the main difference is that the drifting of the perceived centre was allowed in one case but not allowed in the other case. It becomes obvious that the experiments described in this chapter are a generalization of the Haas effect. The generalization was achieved by a sharper criterion.

2.5 Relation to Reality

By electroacoustic means the effect described might be applied to support the voice of a speaker, a singer, or a soloist in a concert (similar to the Haas effect). Wouldn't it be impressive to listen to the smoky voice of a female jazz singer, completely filling the room and surrounding the heads of her audience, yet at a very low intonation, and enveloped by a touch of reverberation? But joking aside, we are not in a cinema and do not want to impose a risk to the ears, exposing them to a sound pressure level which is too high by 25 dB. Furthermore, such applications would require a lot of technical equipment.

The important question is whether the described experiments have any meaning for real concert halls. The first quick answer will probably be negative as there are no such strong early reflections in reality. However, strong or feeble reflections will basically interact in the same constructive manner. Thus the second answer, being more reasonable, must be positive.

Let us discuss the situation, starting with the curves of Fig. 2.10: If the first reflection has a delay time near 15 ms and a relative level near 0 dB it is not much below the first drift threshold. Thus it should have a noticeable effect, and this agrees with practical experience from real concert halls. The second drift threshold is lower in that case, as it is based on a lower level of the first reflection. Thus a second strong reflection at 25 ms, for example, gets a chance to contribute to the image, and so on. However, we must expect that each additional millisecond delay of any early reflection, or a slight lack of power will result in the loss of a nice piece of diffuseness.

It is the opinion of the author that sound reflections from the ceiling do increase the clarity, but do not contribute to the diffuseness. The sound images shown in Fig. 2.11 suggest that any reflection from the ceiling might even *reduce* the diffuseness. It is known that high coherence of the signals received at the ears has an overemphasizing effect in the range above the head [32]. The reader may imagine how the image in Fig. 2.11.6 changes if a high concentration of black draws back to the centre, thus indicating reduced diffuseness.

The methods described might be used for effectively investigating how reflections from the ceiling can in fact contribute to the subjective impressions. Quite generally, the anechoic chamber is the proper tool for exploring the onset of reverberation straight away, either under extreme conditions or near reality. The experiments described suggest further investigations of that kind. The fourth chapter of this book deals with the question how the delays and the levels of the reflections interact in the auditory system.

When explaining the first three partial experiments of the series, we mentioned a few sampling experiments concerning the directional dependence of the drift thresholds. Happily, no dependence on the direction of the reflections could be detected within the measurement accuracy. In terminating the chapter, this strange phenomenon will now be specifically checked once again.

As the directional arrangement of the reflections is of little importance for the drift threshold, the same might be true of the subjective sound image. In order to follow up this point the two partial diagrams of Fig. 2.12 may be compared; these were acquired with groups of 20 and 21 subjects (the left part is a copy of Fig. 2.11.2). The absolute sound pressure level of the direct sound was 68 dB, and only the loudspeakers marked were active. The reflections have the same delay times and nearly equal levels in both cases,

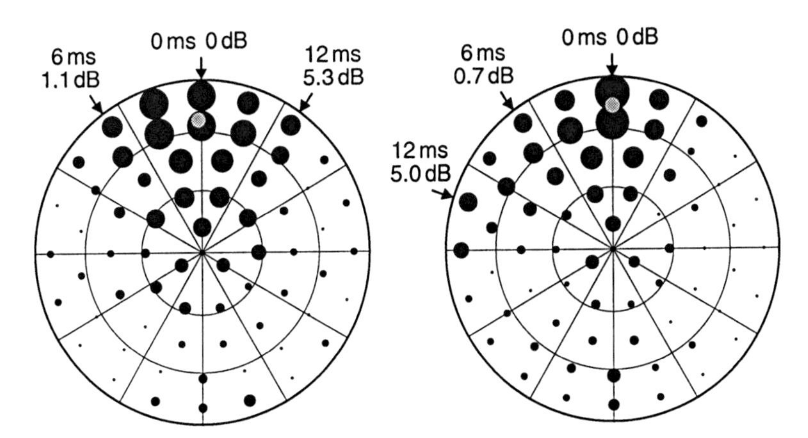

Fig. 2.12. Sound images with symmetric/asymmetric reflections. Centre of the voice: ◍

but the directional arrangements are quite different. In Fig. 2.12.2 there is no reflection from the right.

The right partial diagram, too, shows sound components perceived from above and from behind, and even in the right hemisphere where there is no physical sound at all. Of course, the surprising result must be accepted as it is, and the suspicion that a relevant percentage of the subjects had hearing damage would be out of place (the "resolving power" of a group of subjects may be judged by looking at Fig. 2.3.1). Although the sound was perceived as differently distributed in the two cases of Fig. 2.12, the degree of diffuseness seems to be quite similar. Furthermore, the right partial diagram shows that the azimuthal angle of the perceived centre of the voice *remained at* 0°. Henceforth the stability of the grey spots in Fig. 2.11 appears as more plausible, and the measured drift thresholds are confirmed once again.

The level value of the drift threshold is almost exclusively determined by the delay and the relative level of a reflection, but not by its direction. As an extra feature, a certain degree of diffuseness seems to be given automatically if reflections arrive from *whatever direction*.

How can we understand the drift thresholds to be so strangely independent of the physical sound directions? Might the constant arrival of portions of sound energy be the only thing of importance?

This is a hint for regarding the time functions of the loudness, which are investigated in the fourth chapter. Exciting questions should be treated with moderation, and therefore we may interrupt the discussion of drift thresholds.

3

Definition of Diffuseness

3.1 Introduction

Music, in all its forms, is astonishingly diverse. Indeed, the set of possible compositions might be infinite. When music is presented in a concert hall, the diversity is multiplied by the acoustical effects of the room. Music will certainly defy our attempts to make sweeping generalizations in regard to how it is experienced in various settings.

The symphonic music of a big orchestra is a good example. Such music may exhibit an outrageous number of possibilities. It can generate complex and finely structured impressions in the subjective space of perception. A wide pallet of sound colours, a large dynamical range, a clear control of depth, and rapidly changing directional impressions with varying precision are the gripping elements of such concerts.

Acoustical investigations in this field do demand some thematic restriction. Therefore the *directional multitude* shall be the exclusive topic of this chapter. The main emphasis shall be put on directional impressions which are *widely fanned out* like a "cloud of sound" [32]; pointlike or needlelike directional impressions shall be of less interest.

Cloudlike sound impressions are especially clear in large hall churches. For example, if an organ with 60 registers starts playing J.S. Bach's toccata BWV 565, large decaying clouds of sound can be heard, passing through the church for several seconds. During this reverberation process the subjective sound image has a spontaneously noticeable "spaciousness".

Three dimensions may be assigned to the term spaciousness, which corresponds to the term "space" used in mathematics. To explain the three sound dimensions the reverberation of an organ is a well-suited example. It is widely extended in two dimensions and shows a clearly perceivable depth. On the other hand, wide two-dimensional sound impressions with a lack of depth are basically possible, having no spaciousness in the correct meaning of the word. This was mentioned in the text to Fig. 2.11. In the following the depth of the sound image will be disregarded purposely. Only its two-dimensional spread, in other words its directional multitude which is termed "diffuseness", will be investigated.

3.2 Diffusity versus Diffuseness

The aforementioned 1967 textbook contains the known postulation that physical sound waves envelop the listener on all sides (see Chap. 2). Special acoustic measures are recommended in order to achieve a certain *diffusity* [35, 36]: a subdivision and structuring of boundary surfaces, and even suspended *diffusors*, diverting the sound in various directions. These proposals are based on the tacit assumption that a diffuse *subjective sound image* can be achieved only if the *objective, physical sound* is evenly distributed over all directions (briefly: sound = sound). However, the interaction of the sequenced reflections during the formation of a subjective sound image in the auditory system received no attention in 1967.

The directional distribution of the sound energy in a room may be measured with a rod-shaped directional microphone, or with a microphone installed in a parabolic mirror. The room is acoustically excited in a permanent manner. While a motor is slowly turning the main lobe of the microphone, either horizontally or vertically, the sound pressure level is continuously measured and plotted or stored for later evaluation. After a proposal by R. Thiele [4] *the directional diffusity* is calculated from these data. The result is a number between 0 and 1. The value 1 means a completely even distribution of the sound energy over all directions. Conversely the value 0 means that the sound was received from only one direction, for example from a single loudspeaker in an anechoic chamber.

In order to demonstrate the measured directional distribution of the sound energy, E. Meyer and R. Thiele [5] constructed little hemispheric metal bodies covered with pins of different lengths. At first glance these "spines" of a "hedgehog" show how the sound is distributed (Fig. 3.1). The use of this clever representation method is a reason why the basic idea spread worldwide and has stimulated acoustic research work since 1956. Investigations and measurements concerning diffusity are now in the toolkit of any acoustic consultant working at a newly constructed hall or during structural changes.

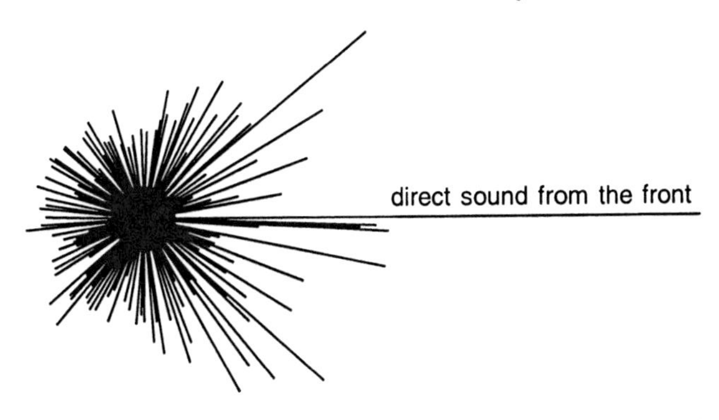

Fig. 3.1. Transcribed photograph of a model hedgehog, showing a high directional diffusity $d = 0.8$ of the sound [5]; top view with perspective distortions

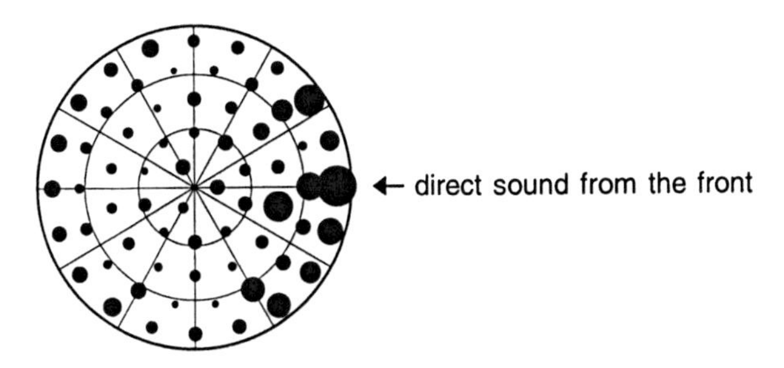

Fig. 3.2. Different form of representation for the model hedgehog of Fig. 3.1

An example of a model hedgehog is shown in Fig. 3.1. The length of each spine represents the sound energy incident from the indicated direction. If printed on paper the model has to be shown in a top view, in a side view, or in any other projection; and this two-dimensional restriction causes distortions in our perspective.

Figure 3.1 shows a top view, allowing judgement of the wall reflections in a room. A side view would be preferred for reflections from the ceiling.

The perspective distortions contained in Fig. 3.1 might be avoided by the form of representation chosen for Fig. 3.2. The diagram is meant to show the same model, but this time the sound energy is represented by the area of the black circles. The original data concerning Fig. 3.1 are no longer available after 50 years. Thus the elevation angles and the proper lengths of the 125 spines shown in Fig. 3.1 had to be estimated or guessed. Therefore Fig. 3.2 can show a form of representation, but it is unsuitable for evaluation, especially as the total number of spines was reduced to 65.

The calculation of diffusity values shall now be explained with the use of a concrete example. For this purpose we select the sound field of Fig. 2.11.6,

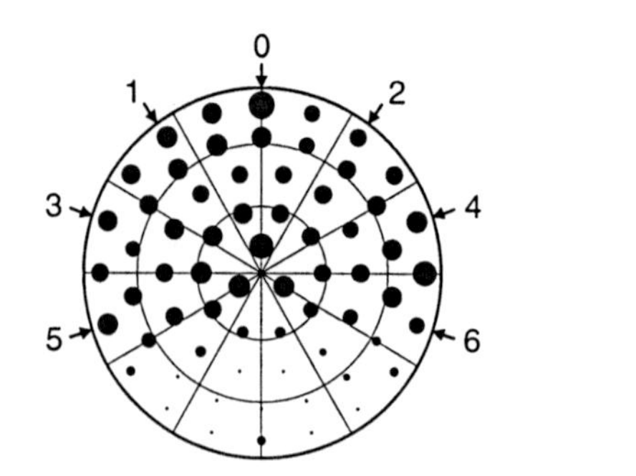

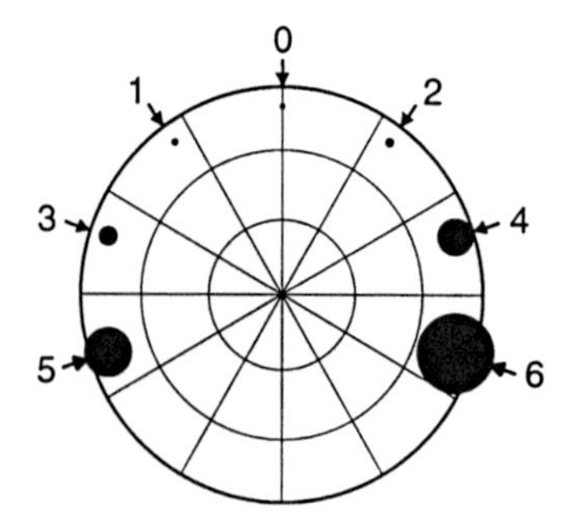

Fig. 3.3. *Left part*: subjective sound distribution (reprint of Fig. 2.11.6). *Right part*: "model hedgehog" showing the energy of the sound components

which is reprinted in Fig. 3.3 (left part), and which is already known to the reader. Thus it allows comparisons and helps to simplify the following discussion.

When judging the sound field the subject's head is placed at the centre of a hemispheric arrangement of 65 loudspeakers. As the subjective sound image covers a great part of the domed surface above the head, the corresponding diffusity value is expected to be rather high.

Seven loudspeakers on the lowest circle were active, as indicated by arrows in the diagrams. These sound components are part of an impulse response with a total duration of 37 ms, representing a simulated onset of reverberation. The physical data of the reflections are summarized in the following table; they are related to the direct sound (spine #0). The values correspond to the first points of the drift thresholds shown in Fig. 2.10.

The area of the black circles in the right part of Fig. 3.3 is proportional to the energy values listed in Table 3.1. The "model hedgehog" shown in the figure is, of course, free from perspective distortion.

The energy values represented by the model hedgehog serve as a basis for determining the directional diffusity d defined by R. Thiele [4, 5]. In this case it has to be calculated with 65 possible sound directions. These directions are given by loudspeakers in the anechoic chamber, evenly distributed on a hemi-

Table 3.1. Physical data of the sound reflections concerning Fig. 3.3

Spine number:	0	1	2	3	4	5	6
Azimuthal angle:	0°	+36°	−36°	+72°	−72°	+108°	−108°
Delay time:	0 ms	6 ms	12 ms	18 ms	24 ms	31 ms	37 ms
Sound press. level:	0 dB	1.1 dB	5.3 dB	12.0 dB	17.7 dB	20.3 dB	24.3 dB
Relative energy:	1	1.3	3.4	15.9	58.9	107.2	269.2

spheric surface. Each of these loudspeakers, averaged statistically, generates the sound energy

$$M = 1/65 \cdot \sum_{n=1}^{65} E_n = 7.03.$$

The averaged absolute deviation from this value is

$$\Delta M = 1/65 \cdot \sum_{n=1}^{65} |E_n - M| = 13.01.$$

With these two values we have to calculate the quotient

$$m = \Delta M / M = 1.85.$$

If only the direct sound is used, the same procedure gives the values

$$M_0 = 1/65, \qquad \Delta M_0 = 1/65 \cdot (|1 - 1/65| + 64 \cdot |0 - 1/65|) = 1.97/65.$$

Once again we have to calculate their quotient

$$m_0 = \Delta M_0 / M_0 = 1.97.$$

Finally, the diffusity d may be calculated:

$$d = 1 - m/m_0 = 1 - 0.94 = 0.06.$$

This small value $d = 0.06$ briefly describes the directional distribution of the sound energy in the given example (see the strange model hedgehog shown in Fig. 3.3). However, 28 subjects established that a large percentage of the hemispheric dome above the head was evenly covered by the sound image (see the left part of Fig. 3.3). *This subjective effect is not adequately described by the value d.*

Once again we are confronted with the question of whether physical data or subjective impressions might be more suitable for representing reality. It was already pointed out that relations between these alternatives have to be sought. However, if a clear decision is unavoidable, acoustical quality means subjective impressions. Therefore, just as for the known *diffusity d*, we shall try to define the *diffuseness D* in order to describe the subjective directional impressions in a sound field.

It should not be forgotten that the intended definition is based on the questioning of subjects, and it is a known fact that research concerning subjective effects is indeed lengthy. However, we must begin just the same, despite the difficulties involved.

The history of science shows that the investigation of an acoustical phenomenon usually begins with purely physical methods. Later on, when sufficient amounts of knowledge have been acquired, properties of the auditory system may be more and more taken into account. The slow scientific acquisition of the term loudness is a good example of this process. Its present precision could be achieved only over a period of more than 100 years. This progress is represented by the following sequence of terms: sound pressure $\rightarrow$ sound pressure level $\rightarrow$ loudness level $\rightarrow$ frequency-weighted sound pressure level $\rightarrow$ loudness. The respective units of measurement are: Pa (pascal), dB (decibel), phon, dB (A/B/C), sone.

The last quantity mentioned in the series is the loudness, measured in the unit sone. This quantity contains a vagueness which will be relevant in the following and has to be explained. The term loudness ends with the syllable "ness", which is meant to indicate an impression of the senses. On the other hand, there is the loudness analyzer which is a physical device [37].

It measures the loudness in an analog or digital manner; for example, 4.2 sone. Thus we might confuse an impression of the senses with a numerical reading as *both of them* are called loudness. The sentence "I can't stand the loudness in the railway station" does not mean that somebody cannot stand a numerical value, but it means that he or she cannot stand the loudness.

The quick use of one word in two senses might indeed be confusing. However, as concerns the loudness, investigations with large groups of subjects showed that a proper conjugation number $\leftrightarrow$ impression is definitely possible [37].

An ambiguous use of a word may even be helpful if misunderstandings are prevented by the context. The ambiguity is a classic means for the construction of terms and allows attaching a short label to a complicated fact. In this chapter the term diffuseness will be used for the directional multitude in subjective sound impressions. Second, the same word will be used for a numerical value which signifies the degree of this multitude.

3.3 Computer Program for Defining Diffuseness

Here and in other parts of the book, the treatment of details already mentioned in a different relation is unavoidable. Let us hope that this is more helpful than disturbing!

As discussed in Chap. 2 the diffuseness of a sound field can be determined by a group of subjects. Each of them has to note down the sound directions appearing in his or her subjective space of perception. This ambitious task should be restricted to a clearly defined sound probe; for example, to a few bars of a piece of music, a chord of an organ, continuous speech of a single talker, or noise pulses. In some cases the physical sound directions and the perceived directions have little in common; even drastic contradictions are possible. However, *the objective physical sound directions are not asked for at all.* They even have to be unknown to the subjects in order to avoid confusion. Thus the subjects are asked not to move their heads during the test. In particular they are not allowed to leave the seat to find out the "real sound sources". In short, they are asked only for the perceived directions.

The following considerations are restricted to the upper hemispheric space, that is, to the subjective space of perception, but only above the listener's head. Curious listening impressions like trumpet sounds perceived inside the head, or bass sounds perceived "from below the carpet" shall be ignored. In order to note down the perceived directions, a circular plan of the upper hemispheric space is handed to the subject. In fact the plan is an equi-areal projection on the horizontal plane, just like a celestial chart in an atlas. In this plan the perceived directions, like the stars in the sky, have to be marked. Let us hope that the subjects are able to associate the plan with their subjective space of perception. Their imagination is supported by the fact that about 20 loudspeakers of the hemispheric arrangement are permanently visible.

The area of the circular plan is subdivided into equal parts. The fineness of the subdivision depends upon, among other things, the time available to the subjects and on their experience with extensive listening experiments. In principle a very fine raster as well as many intelligent subjects showing an unlimited readiness for cooperation would be desirable. In the experiments described in Chap. 2 there were 65 fixed directions which had to be either marked or not marked. In each of the sound fields presented each subject would thus actually have to make 65 yes–no decisions. It would be unrealistic to hope for a working style which is always that careful! Yet, as an example, the following considerations are based on the aforementioned plan with 65 directions in order to maintain the comparability with Chap. 2.

Let us now carry out a gedanken experiment. We assume that two different sound fields had to be judged by three equally large groups of subjects. Their hypothetical listening results are shown in Fig. 3.4. As mentioned in Chap. 2 the area of each little black circle represents the percentage of the subjects who marked the respective direction. Which diffuseness values shall we assign to these three partial figures?

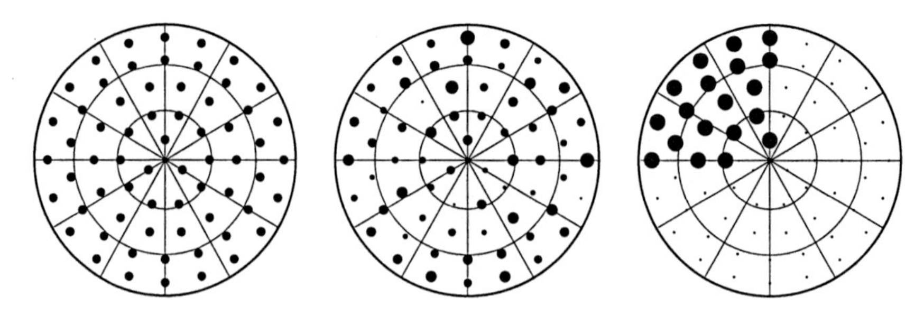

Fig. 3.4. Hypothetical listening results with two different sound fields

The examples are intentionally constructed to contain exactly the same total number of marks and the same amount of printer's ink in each case. The first group of subjects judged the first sound field to be perfectly diffuse, as all directions were marked, with no detectable weight difference. This extreme agreement within the group of subjects seems to be hypothetical, and it is so. Anyhow, this case has to be assigned the diffuseness value $D = 100\%$.

The second hypothetical group of subjects had to judge the same sound field. However, these subjects' attention spans, their patience or enthusiasm for the task, or perhaps even their hearing capabilities appear to be reduced in comparison with the first group. In this case only 62 out of 65 directions were marked with different percentages. Thus, the second partial diagram appears less contrived than the first one, as statistical differences have to be expected in the results of 20 subjects. Yet these subjects, too, determined "somehow" that the sound field generated the diffuseness $D \approx 100\%$. However, this fact is not caused by the exact equality of the total numbers of marks contained in the two figures! The decisive reasons for a high diffuseness value are that the marks are not clearly concentrated in certain regions of the plan and, by the same token, there are no empty regions.

The third partial diagram, too, contains neither more nor fewer answer marks than the other partial diagrams. However, in this case the answer marks are concentrated in a small region where they are uniformly distributed. We'll construct a mathematical procedure to evaluate the diagrams, and we shall assign a diffuseness value near 25% to the third partial diagram. The value might be a little smaller as well, because the answers are concentrated in a sector that begins to resemble one limited spot.

At this point we have to construct a mathematical procedure which generates "reasonable" diffuseness values in any case, especially in the examples explained above. To solve this problem, a computer program was designed using the language BASIC. The program was continually improved over several

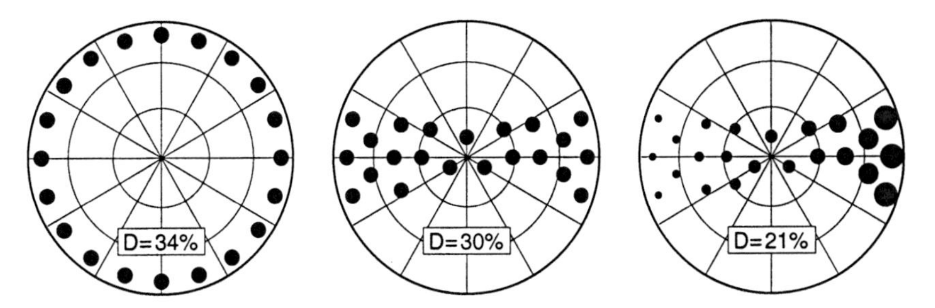

Fig. 3.5. Diffuseness values with hypothetical listening results, and with the same total number of marks (amount of printer's ink) in each case

versions. When applying the program to actual listening results, the insufficiencies, contradictions, and even more the program errors, become obvious immediately. We tested the program by applying it to the sound fields explored in Chap. 2, and those tests required about five minutes. Even if we achieve only plausible numbers with these sound fields, many hypothetical and extreme cases have to be carefully checked, too. Some examples are shown in Fig. 3.5.

The main problem during this development process was, of course, that the diffuseness was not defined in advance. However, its precise definition must be worked out in this manner. In other words, there were no definitely correct diffuseness values while the program was being worked out. A first program version, announced at a symposium in Japan in 1995, proved to be unnecessarily complicated [38].

Some important rules were observed when the program was developed. They relate to the circular diagrams with 65 given directions:

- If all sizes of the black circles contained in a diagram are changed in the same proportion, the diffuseness value shall not be influenced.
- The diffuseness lies in the mutual relations of the black circles.
- The contribution of any two circles to the diffuseness increases with the angle between their directions, or with their spacing in the plan.
- The contribution of any two circles to the diffuseness increases with the geometrical average of their respective sizes.

The following comprehensible procedure was finally selected:

```
'-------- program defining the diffuseness (excerpt) ---------
'based on perceived sound directions logged up by ≥20 subjects
'
'A(I),E(I) = azimuthal and elevation angle of the direction I
'A(K),E(K) = azimuthal and elevation angle of the direction K
```

```
'the directions I,K put up a difference angle D(I,K) =
'ARCCOS(COS(E(I))*COS(E(K))*COS(A(I)-A(K))+SIN(E(I))*SIN(E(K)))
'
'P(I) = % of subjects who perceived sound in the direction I
'P(K) = % of subjects who perceived sound in the direction K
'Average = percentage averaged over all 65 directions
   Psum=0: FOR I=1 TO 65:Psum=Psum+P(I): NEXT :Average=Psum/65
'
'Prmax = largest product of percentages in the distribution
   FOR I=1 TO 65:P_(I)=P(I): NEXT : SORT P_(0)
   Prmax=P_(65)*P_(64): IF P_(64)=0 THEN Prmax=P_(65)*P_(65)
'---------------------------------------------------------------
Homogen=0:Diffuse=0' calc. diffuseness with directional pairs
FOR I=1 TO 65: FOR K=I+1 TO 65
  Homogen=Homogen+ Average*(1+D(I,K)/45)' reference value
  IF P(I)*P(K)>.05*Prmax THEN '        limit of relevance
     Diffuse=Diffuse+ SQR(P(I)*P(K))*(1+D(I,K)/45) ENDIF
  NEXT K: NEXT I
D=Diffuse/Homogen:PRINT USING "##.###";" Diffuseness D=";D;
'---------------------------------------------------------------
```

The diffuseness value was calculated as a sum of (many) contributions. However, they were taken into account only if the product `P(I)*P(K)` was greater than 5% of the largest product occurring. The angle put up by two directions, or the distance of the corresponding points in the plan, was represented by the weight factor `(1+D(I,K)/45)`. Thus, the weight of the largest possible angles was more than three times the weight of the smallest.

If applied to Fig. 3.4, the program calculates the three diffuseness values $D = 100\%, 94\%, 21\%$. The value 100% describes the homogeneous distribution of black circles with equal sizes. The second partial figure with an irregular distribution is assigned the diffuseness $D = 94\%$. Thus the ideal value 100% was not reduced much by the inaccuracies purposely constructed. A value $D \approx 25\%$ was expected for the third partial diagram, and it was pointed out that the result might be slightly smaller. These expectations are fulfilled by the program.

If the sound image contains just one single spot, the program calculates the diffuseness value $D = 0$ in any case. It is uncertain whether a group of subjects would ever produce such a definite result. For discussing this case we may consider Fig. 2.3.1 (page 49) once again, which must be regarded as a general reference. The diagram was achieved with 21 subjects working one by one in the anechoic chamber. *Only a single loudspeaker*, placed exactly in front of the subject, was emitting continuous anechoic speech. However, apart from the physical sound direction, the subjects marked several other directions in which they had perceived sound! In this fundamental case, which may be recommended for further experiments, the computer program calculated the diffuseness value $D = 4.5\%$. It may be assumed that the result from a group of inexperienced subjects cannot fall drastically below this value.

3.4 Application to the Onset of Reverberation

The diffuseness, expressed concisely, is the multitude of perceived sound directions. The description of this multitude by one numerical value D requires data compression. In principle, the calculation might be based on data acquired by a single excellent subject. However, the subject would have to judge

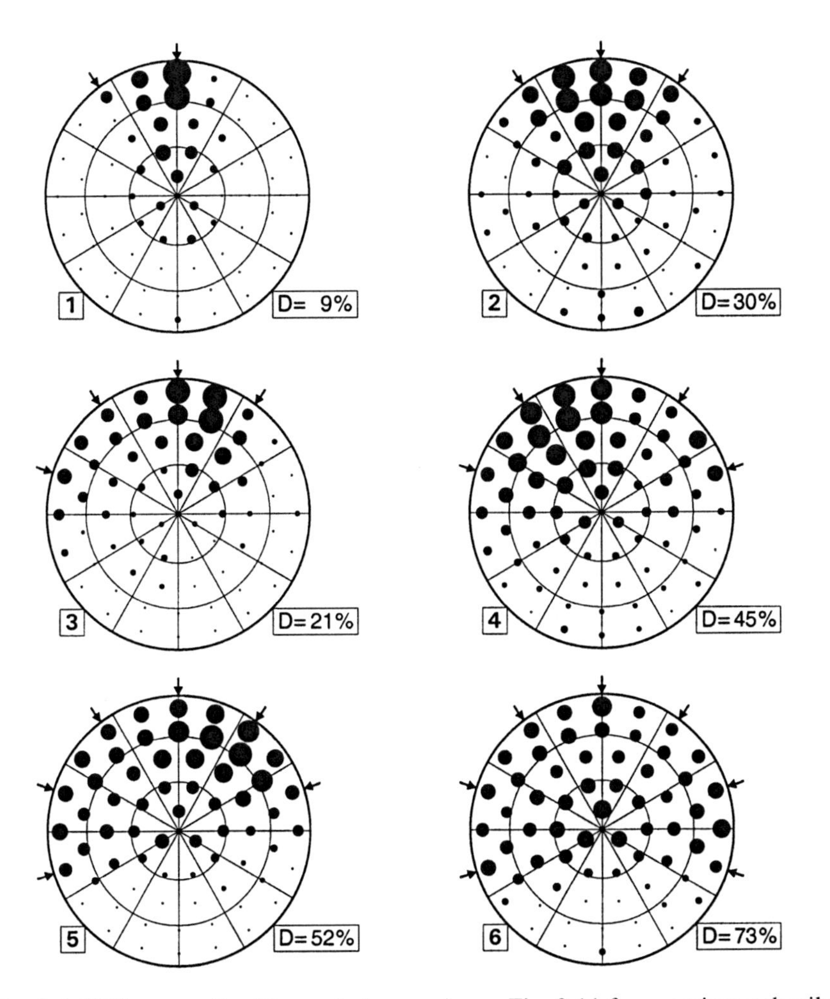

Fig. 3.6. Diffuseness D with anechoic speech, see Fig. 2.11 for experiment details

and to log the relative loudness of each perceived sound component because these data are needed for the weight factors. In Fig. 3.6 the situation is less difficult. These diagrams were copied from Chap. 2 (Fig. 2.11) and each of them is based on the listening results of at least 20 subjects. Thus the weighting is given automatically, as the directional components were noticed and logged by different percentages of the subjects. The six diffuseness values in Fig. 3.6 were calculated with the definition given by the computer program described.

The reader is invited to examine the six diagrams one by one and to describe the directional multitude expressed in each case by a percentage value between 0 and 100%. He or she must be advised, however, to avoid long considerations. It can be assumed that these estimated values come close to the values calculated by the computer. In short, the calculated values seem to be plausible and correspond with experience. Once again we may recall the fact that these sound images show a lack of depth, even at the high diffuseness $D = 73\%$. Therefore the impression of diffuseness is insufficient for genuine spaciousness.

3.5 Simplified Measuring Scheme

In each of the examples described here, at least 20 subjects were requested to log the perceived sound directions in the circular plan. It was not possible to perform a systematic analysis for determining how the subjects solved the problem of noting down a subjective impression on paper. However, with regard to each single questionnaire a variety of working styles or hearing experiences is discernible. It becomes quite obvious that in a group of 20 subjects some are so skillful and sensitive, or so intelligent, that their personal record comes close to the average result for the group. Therefore it is imaginable that diffuseness values of common relevance might be determined by only a few experienced subjects, or even by a single person.

As mentioned earlier, the subjects would have to note down the perceived directions including their relative loudness values. In a simple sound field the total number of perceived directions may be small enough for comparison of all the loudnesses. However, mutually comparing the loudnesses of 50 or 60 directional components in a complex sound field and noting down that many percentage numbers appears to be an unreasonable task, indeed.

Let us have another look on the little circular diagrams in Fig. 3.6; see, for example, Fig. 3.6.5. The largest black circle in this diagram is surrounded by seven smaller circles, but their areas are not extremely different from that

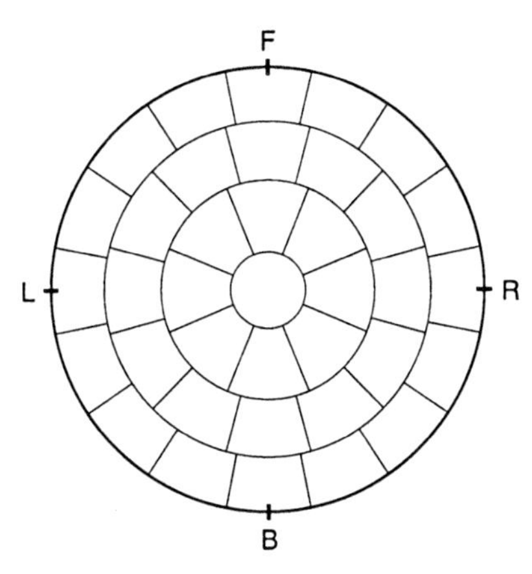

Fig. 3.7. Simplified scheme for perceived sound directions and loudnesses

of the largest circle. This moderate transition leads to the conjecture that the diffuseness might as well be measured with a smaller number of directions on the questionnaire. The conjecture is supported by the aforementioned fact that a needlelike sound from the front generates a diffuseness value of $D = 4.5\%$. This value corresponds to a rather low resolving power of a group of inexperienced subjects.

The simplified scheme proposed in Fig. 3.7 shows 37 sections of equal size which indicate the same number of spatial directions. It should be the subject's first task to determine the direction of greatest loudness and to mark the corresponding field. In relation to that sound component, the loudnesses of the other components have to be estimated and noted down as percentage numbers in the scheme. Even crosses in a few fields would be sufficient for calculating diffuseness values with the computer program adapted to 37 directions.

In order to test the applicability to an example, the data in Fig. 3.6.5 may be transcribed to the simplified scheme. In Fig. 3.6.5 the greatest loudness comes from the front right, and thus 100% has to be entered in the corresponding field of Fig. 3.7. The percentage for each of the other fields in the scheme is determined by estimating and summarizing the sizes of several black circles in Fig. 3.6.5, always in relation to the field of greatest loudness. The computer uses these numbers to calculate the diffuseness. Results in the range $D = 50–54\%$ were achieved with slightly different sets of estimated data. If the scheme has only 25 directions, instead of 37, the situation is quite similar. In this case, however, the redundancy is at the lowest acceptable limit.

The reduced number of directions makes the listening task much less difficult, and it is even less difficult if the scheme contains only 25 fields. This opens up the possibility of measuring diffuseness values using only stereo equipment in a living room. With a modern CD player, short musical sections may be continuously repeated until the scheme is completely filled in. Such diffuseness values might be helpful for comparing different recordings, or for judging the quality of the stereo equipment or the room itself.

In concert hall investigations the measurement of diffuseness is of great importance, and in order to avoid low redundancy the evaluation scheme should have more than 25 fields. Here, too, the measurements may be carried out by a single subject. However, for collecting information about the entire hall it will be more advantageous to employ several subjects at the same time; the simplified scheme of Fig. 3.7 is explained and handed to each of them. As with the known syllable-comprehension tests, the subjects may be distributed in different parts of the hall. Even concertgoers might take part in such measurements. The subjects are requested to log the sound directions perceived during the onset of reverberation, and for this purpose a well-instructed orchestra repeats a few bars of a musical piece again and again, perhaps just one chord. We can also imagine using loudspeakers to excite the hall. If the genuine reverberation is not the topic of investigation, as in this chapter, it must be ignored as far as possible. As an alternative, the onset of reverberation of the hall under investigation may be simulated in an anechoic chamber, excluding the genuine reverberation.

The future will show which of these procedures is suitable for the proposed measurements.

The definition given in this chapter suggests that diffuseness values cannot be determined without the use of listening results from living persons. However, a quite different approach, based on time functions of the loudness, is offered in Chap. 5. In the onset of reverberation these functions allow a distinction between the effect of the direct sound on the one hand, and the diffuseness generated by the early reflections on the other hand.

4

Theory of Drift Thresholds

4.1 Law of the First Wave Front

This chapter's aim is to interpret the experimental results concerning the onset of reverberation presented in Chap. 2. However, the measurement of diffuseness treated in Chaps. 2 and 3 shall also be considered.

Let us recall the definition of the onset of reverberation in a concert hall. When determining the impulse response, the room is excited by a shot. If the signal received by a microphone is displayed on the screen of an oscilloscope, a series of many needlelike lines is visible and ends at 2 s, for example. The first impulse and its subsequent impulses, up to a delay time of about 50 ms, form the onset of reverberation. These impulses correspond to the direct sound and the early reflections. In considering the oscillogram, the physicist will ask himself about the subjective sound effects caused by these needlelike peaks visible on the screen. How is the series of early reflections evaluated by the auditory system? Which sound effects, above all which directional impressions, must be expected?

The following famous sentence can be found in a textbook on acoustics [31]: "Finally ... it may be mentioned that, concerning the directional impression, *only the first wave front* is evaluated by the ear if the signal, shifted in time, arrives from several directions." The clarity of this statement about a complicated matter led to the formulation of the "law of the first wave front", and it contributed to the fact that the sentence is quoted rather often. However, the absoluteness of the formulation provokes doubts about its validity.

The doubts may be supported by Fig. 2.3 (page 49), which shows listening results concerning the Haas effect. In all three cases described in this figure, the first wave front is defined by the direct sound arriving from the front (azimuthal angle 0°). However, the ear notices *more than this single direction.* Instead, a multitude of directions is perceived simultaneously. As can be seen from Fig. 2.3.2 and 2.3.3, the directions with a large weight in the sound image may in fact deviate from the line of view. Moreover, the perceived centre of the voice (grey spot) is by no means clearly linked to the first wave front. In Fig. 2.3.3 there are even *two separate centres* of the voice marked by the subjects. The reason is a 50 : 50 split of the statistical distribution of the answers, indicating an uncertainty in the auditory system. Thus the law of the first wave front is valid only within certain limits, which were disregarded in the case of the Haas experiment.

The listening results presented in the six partial diagrams of Fig. 2.11 (page 59) come much closer to the situation in concert halls. In this experimental series the perceived position of the talker (grey spot) always *remains near the line of view*, which is the direction of the first wave front. However, the levels of the six reflections form a rising series, beginning with small steps, and it must be admitted that this rising tendency is not typical for concert halls.

In these experiments too, the subjective sound image was not restricted to just one precise direction. Instead, from step to step in the experimental series, the directional distribution spreads more and more widely, and a main direction can be detected less and less clearly. Finally, if Fig. 2.11.6 is considered, it becomes obvious that a diffuse sound impression may as well contain *no main direction at all*. The grey spot seems to indicate such a definite main direction, but it represents the statistical average of many individual answers, and these answers showed a large fluctuation in the last experiment.

If a dominating direction cannot be made out in the image, the law of the first wave front has lost its meaning. The author admits that he has purposely tried "to evade the law".

But back to the question asked at the beginning: Which methods are applied by the auditory system to evaluate the impulse response of a room, especially the series of early reflections? The acoustical analysis carried out by the ear has many different aspects, and we do know that this data processing can be carried out only at a limited speed. This fundamental limitation is contained in the law of the first wave front, too. Thus, the ear works with a certain slowness while permanently calculating a weighted average. Expressed mathematically, this means a sliding integration, putting the largest weight on actual events in the present. However, the farther back an event lies in the past, the less it will be contained in the actual average. Such calculations may be carried out perfectly by a digital computer or an analog circuit.

The first part of an impulse response of a room, concerning the onset of reverberation, is shown in Fig. 4.1. The diagram was worked out with a room

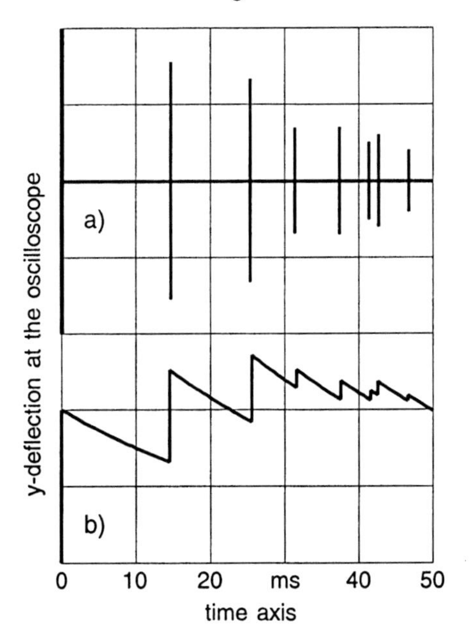

Fig. 4.1. Impulse response of a narrow and high cuboidic room, **a** original microphone signal, **b** processed by an RC-integration

shaped like a narrow and high cuboid, comparable with Boston Symphony Hall [33], but scaled down by a factor of 2/3.

An original impulse response as shown in Fig. 4.1a does not include any property of the human ear. As mentioned earlier, this impedes the interpretation. Therefore, following a proposal made by H. Niese, the microphone signal is rectified by a square-law diode and then conducted to an RC-integrator at the input to the oscilloscope. The time constant RC (resistor, capacitor) is adjusted to 28 ms, in accordance with the slowness of the ear [39]. With this preliminary model for evaluating the loudness in the human ear, we calculated the time function shown in Fig. 4.1b. It explains the superposition of loudnesses caused by the direct sound and the early reflections after a shot. Except for the initial rise the steps in part (b) are smaller than the amplitudes of the corresponding pulses in (a). This is caused by the square characteristic of the rectifier.

4.2 Calculating Time Functions of the Loudness

If the room is excited by speech instead of a shot, the RC-integrator will generate smooth curves. In principle these curves might be sufficient for allowing a superficial interpretation of the drift thresholds presented in Chap. 2. However, a precise loudness model realized on the computer will be much more suitable (p. 84). It is based on papers published by E. Zwicker and A. Vogel [6, 7, 37, 40] and requires some explanation.

In the auditory system, any signal received at the ear will immediately be decomposed in an adequate number of filters briefly called "frequency groups", with a bandwidth of about $1/3$ of an octave each. They are centred in accordance with the signal and are formed along the basilar membrane in the inner ear. The output signals from the filters are then processed in five steps (a to e) explained in the computer program detailed in this chapter. This results in partial loudnesses, their sum being the loudness.

Fortunately, the complicated spectral decomposition is unnecessary for the interpretation of the drift thresholds. In the computer program the spectral decomposition is therefore replaced by an additional level value called a "spectral extra". This value expresses the fact that a signal appears to be louder, at one and the same sound pressure level, if its energy is distributed over more than one frequency group. The procedure of determining the spectral extra shall now be explained with an example.

Let us assume the energy of the speech to be concentrated in only two adjacent frequency groups centred at 630 Hz and 800 Hz, with a sound pressure level of 62 dB in each of the groups. Thus a wide-band root-mean-square meter, adding up the incoherent signal components, would show 65 dB. In order to determine the loudness with the aid of a Zwicker diagram [40], we have to start by selecting the horizontal line at 62 dB in each of the frequency groups. These two lines define an upright, nearly rectangular area which has to be enlarged by the "masking slope" of the upper frequency group. The overall area enclosed by these lines determines the loudness level 67.5 dB to be read on a special scale. The extra 2.5 dB estimated in this way includes the A-weighting according to the known equal loudness contours.

In the original version of the loudness model, the loudness N is assumed to be proportional to the square root of the sound pressure: $N \sim p^{0.5}$ [6, 7], and the time constant of the required intermediate storage is given as $\mathrm{RC} = 35$ ms. However, the program described in this chapter uses the relation $N = 1/16 \cdot p^{0.602}$, which conforms with the ISO standard. As a consequence, the intermediate storage (program step c) has to be made a little faster with $\mathrm{RC} = 28$ ms. Otherwise we would have to accept some inaccuracies in the interpretation of the drift thresholds. Regarding this point, RC-values of 28–35 ms are mentioned in the literature.

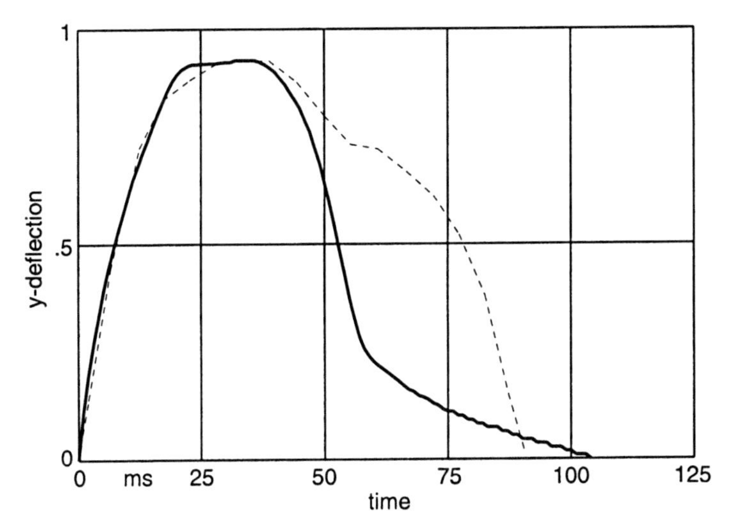

Fig. 4.2. - - - vowel "u", spoken as in "but", +/− envelopes averaged [1]; — shortened "u", replacing continuous speech in loudness calculations

The measurements of the drift thresholds (Chap. 2) were carried out with continuous anechoic speech from a tape recorder. However, time functions of the loudness of continuous speech show a considerable and unforeseeably varying modulation. This hinders the detection of interrelations, and a way out of this awkward situation had to be found.

In the course of the threshold measurements, when listening in a concentrated manner, a curious and important observation could be made: *The drift thresholds are almost exclusively determined by the powerful vowel "u", spoken as in "but"*, appearing again and again in any continuous text read aloud. Observing other sounds or sequences of sounds requires more concentration, and the subject needs more time for decisions.

For the interpretation of the drift thresholds it was therefore attempted to reduce the continuous speech to just one short impulse which is shown by the boldface curve in Fig. 4.2. The envelope of the vowel "u", spoken as in "but", is shown for comparison (dashed line). An oscillogram of this vowel proves the two envelopes of its positive or negative part to be rather different [1]; thus the envelope shown in Fig. 4.2 is an average.

When the oscillogram was stored, just one separate word had been spoken: the German word "zack" which sounds similar to "but". However, the measurements of the drift thresholds were carried out with continuous speech. This is made up of many short, pulse-like sounds, and the vowels, too, may be assumed to be spoken somewhat shorter than in a separate word. Therefore a pulse duration of about 50 ms (boldface curve in Fig. 4.2) was used for the interpretation of the drift thresholds.

Right at the start (`-Begin`) the program for the calculation of time functions of the loudness asks for the level of the direct sound. It should be entered correctly; this means the value used for the measurement of the drift threshold. One must aim at an accuracy of about ±3 dB.

```
'************************************************************************
' BASIC program for time functions of the loudness (excerpt)
' based on B( ),But( ):  short version of the vowel "u"(but)
'************************************************************************
CLEAR: CLS: DEFSNG "A,B,C,D,E,N,P,V": DIM T(10),Db(10),Amp(10)
DIM B(161),But(161),N(161),Nzero(161),Nplus(161)
'---------------------------------------------------- delay times in ms
T1=6:T2=12:T3=18:T4=24:T5=31:T6=37'                   fixed reflections
T(1)=6:T(2)=12:T(3)=18:T(4)=24:T(5)=31'            variable reflections
T(6)=37:T(7)=43:T(8)=55:T(9)=78:T(10)=108
'----------------------------------------------------------------------
'FOR T=1 TO 140:B(T)=1.: NEXT '          square 140ms for calibration
 FOR I=1 TO 105: READ B(I): NEXT '        shortened "u", uncalibrated
-Begin: Vari=0: Out$=" ": PRINT  "Direct sound: dB"
 INPUT @(0,14);Db0:Db0=Db0+3'                     +3dB spectral extra
 P=10^(Db0/20):Calib=.006125*P'             square 140ms,40dB ==> N=1
 FOR T=1 TO 160:But(T)=Calib*B(T): NEXT'         short "u", calibrated
'----------------------------------------------------------------------
INPUT "Reflection: ";Ref
FOR S=Ref TO 10'                      S = delay = spot on the time axis
-Again:Clear5: USING : FOR T=0 TO 160:N(T)=0: NEXT '          steps 1ms
 IF Vari=0 THEN PRINT "Working on basic curve...";
 IF Out$="c" OR Out$="a" THEN PRINT "Working at spot";S;"...";
 IF Out$="s" THEN ' on screen: variable refl.level --> dev.angle
    PRINT @(5,14);"(0=end)";
    INPUT @(5,0);"Spot";S: PRINT @(5,14);"  ";
    IF S>0 THEN S= MAX(Ref,S):S= MIN(S,10) ELSE STOP
    INPUT @(6,0);"Additional level, dB: ";Vari:Clear5
    PRINT "Working at spot";S;", addit.level";Vari;"dB...";
    Vari=10^(Vari/20) ENDIF '         additional level now as a factor
'************** calculating N in steps a,b,c,d,e ********************
'a) ---------------------- total sound pressure = SQR(sum of energies)
SELECT Ref'=Reflection, Db( )=respect. drift threshold of Ref
CASE 1' Ref 1 variable, at the spots S=1...10 of time axis
  Db(1)=1.1:Db(2)=1.9:Db(3)=1.5:Db(4)=1.5:Db(5)=-.8' <--Fig.2.10
  Db(6)=-.9:Db(7)=-6.7:Db(8)=-10.5:Db(9)=-18.5:Db(10)=-27.1
  FOR I=1 TO 10:Amp(I)=10^(Db(I)/20): NEXT ' amplitudes
  FOR T=1 TO 160:N(T)=But(T)^2' sum of energies
    IF T-T(S)>0 THEN N(T)=N(T)+(Vari*Amp(S)*But(T-T(S)))^2
    N(T)= SQR(N(T)): NEXT ' here N(T) means sound pressure
CASE 2' Ref 1 constant, Ref 2 variable at the spots S=2...10
  Db(1)=1.1:Db(2)=5.3:Db(3)=5.2:Db(4)=5:Db(5)=3.1' <--Fig.2.10
  Db(6)=3.7:Db(7)=-.1:Db(8)=-5.6:Db(9)=-12.8:Db(10)=-22.7
  FOR I=1 TO 10:Amp(I)=10^(Db(I)/20): NEXT ' amplitudes
```

```
  FOR T=1 TO 160:N(T)=But(T)^2' sum of energies
    IF T-T1>0 THEN N(T)=N(T)+(Amp(1)*But(T-T1))^2
    IF T-T(S)>0 THEN N(T)=N(T)+(Vari*Amp(S)*But(T-T(S)))^2
    N(T)= SQR(N(T)): NEXT ' here N(T) means sound pressure
'CASE 3,4,5,6 respectively, see Appendix
END_SELECT
'b) ------------------------------------------ fast integrator RC=1.4ms
FOR T=160 TO 1 STEP -1: FOR J=1 TO T-1
    N(T)=N(T)+N(T-J)*EXP(-J/1.4): NEXT J:N(T)=N(T)/1.4: NEXT T
'c) -------------------------------- loudness according to ISO standard
FOR T=1 TO 160:N(T)=N(T).60206:NEXT ' ISO-exponent=2*LOG10(2)
'd) ------------------------------------------- diode and RC-store 28ms
'          Vogel Fig.9 RC=45ms, in the text 35ms. Niese 28ms with p^0.6
E= EXP(-1/28): FOR T=1 TO 159:N(T+1)= MAX(N(T+1),N(T)*E): NEXT
'e) ---------------------------- final integrator, 3x low-pass RC=16ms
FOR T=160 TO 1 STEP -1: FOR J=1 TO T-1
    N(T)=N(T)+N(T-J)* EXP(-J/16): NEXT J:N(T)=N(T)/16: NEXT T
FOR T=160 TO 1 STEP -1: FOR J=1 TO T-1
    N(T)=N(T)+N(T-J)* EXP(-J/16): NEXT J:N(T)=N(T)/16: NEXT T
FOR T=160 TO 1 STEP -1: FOR J=1 TO T-1
    N(T)=N(T)+N(T-J)* EXP(-J/16): NEXT J:N(T)=N(T)/16: NEXT T
'-------------- loudness N(T) ready, store curve ---------------------
IF Vari=0 THEN FOR T=1 TO 160:Nzero(T)=N(T): NEXT' basic curve
IF Vari>0 THEN FOR T=1 TO 160:Nplus(T)=N(T): NEXT' +reflection
'printing commands and other commands follow, see Appendix
'example: basic\curve_ready --> test_refl_on, Vari=1: GOTO Again
```

The program offers three different selections for the output: First, the precise data of the curves can be listed. This makes 45 printed pages for altogether 45 measured points of the six drift threshold curves presented in Fig. 2.10. Second, a list can be printed for each of these curves, showing the influence of the test reflection on the respective basic loudness curve. Third, this influence can be carefully studied on the computer screen while varying the level of the test reflection.

4.3 Loudness and Drift Thresholds

As mentioned earlier, the immensely varying loudness modulation of continuous speech confuses any interpretation. All investigations presented here concerning the onset of reverberation are, therefore, based on the shortened envelope of the vowel "u" (as in "but"). Its time function of the loudness, valid for an absolute sound pressure level of about 68 dB and for the direct sound only, is shown by the basic curve in Fig. 4.3.

The oscillogram of the vowel "u" (as in "but") and also the shortened version of the envelope show an immediate, steep rise which terminates after 20 ms (Fig. 4.2). However, *the loudness* of the direct sound reaches its max-

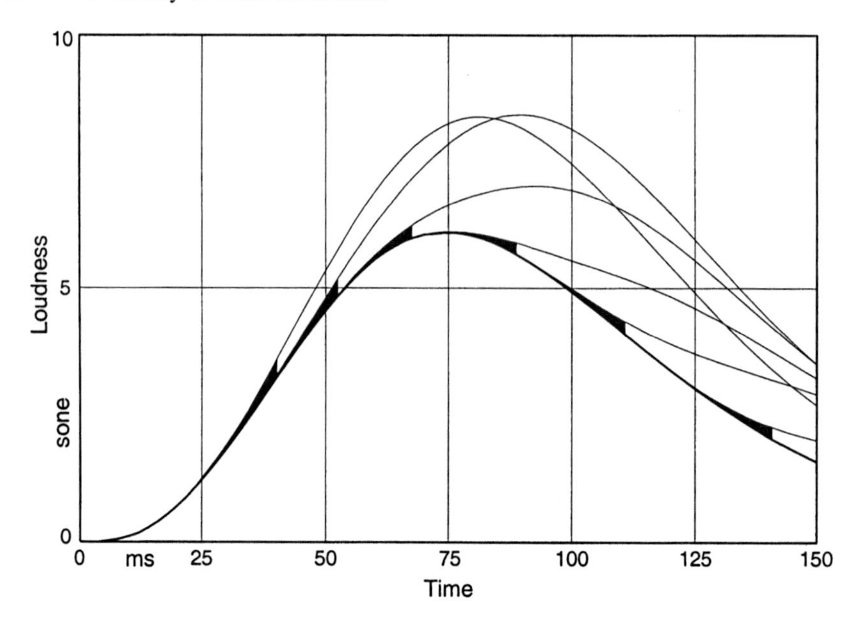

Fig. 4.3. Loudness of the vowel "u" (as in "but"). *Basic curve*: direct sound. *Thin curves*: one reflection added, its level adjusted to the drift threshold

Table 4.1. Physical data of the first reflection concerning Fig. 4.3

Delay time:	12 ms	24 ms	37 ms	55 ms	78 ms	108 ms
s.p. level:	+1.9 dB	+1.5 dB	−0.9 dB	−10.5 dB	−18.5 dB	−27.1 dB

imum only after 75 ms (boldface line in Fig. 4.3). Thus, the loudness rises less fast and fades away even more slowly, which is a clear illustration of the general slowness of the ear.

If a single reflection is added to the direct sound we have to consider the total loudness represented by the thin curves in Fig. 4.3. In this case the height of the maximum and its position in the diagram depend on the delay time and on the level of the reflection. These data are listed in Table 4.1 and are always related to the direct sound. The data correspond to the six deviating curves in succession.

These coordinates correspond to six points on the drift threshold of the first reflection,shown as the lowest curve in Fig. 2.10 (page 58). But back to Fig. 4.3: *The six curves branching off from the boldface basic curve have in common that they belong to one and the same drift threshold curve*. Therefore we might find graphical similarities. The black "wedges" between the curves are intended to indicate the following hypothesis: The reflection has reached the drift threshold if the curve of the total loudness branches off from the basic curve to a precisely determined minimum degree.

The calculation of an angle between the loudness curves, at a certain spot shortly after the beginning of branching, is less difficult than the comparison of two curvatures. For this reason the simpler procedure was actually chosen, and the angles are calculated at a spot 23 ms after the beginning of branching. This fixing proved to be advantageous in a few preliminary calculations. To calculate the angles, the output section of the computer program receives a few more lines:

```
'gradient without/with reflection, 23ms after branching begins
'angles graphically correct only with scale        10ms<->1sone
Nptzero=(Nzero(T(S)+24)-Nzero(T(S)+22))*5'=(difference/2ms)*10
Nptplus=(Nplus(T(S)+24)-Nplus(T(S)+22))*5'=(difference/2ms)*10
Tangdelta!=(Nptplus-Nptzero)/(1+Nptplus*Nptzero)' tang≈angle
```

A conversion tangent → angle is dispensable as all calculated angles are $\leq 10°$. The program will now be applied to the drift thresholds plotted in Fig. 2.10. The measurements were all carried out with a test reflection added to the direct sound and several preceding, fixed reflections (see Chap. 2). The upper curve of Fig. 2.10 is valid for the final experiment of the series. It was carried out with the direct sound, five fixed reflections, and the test reflection. Thus, in this case, there are altogether seven sound components at each of the measured points.

In the computer program the six threshold curves of Fig. 2.10 are treated one by one. For each of these curves the fixed sound components are first combined to calculate the basic loudness curve. Then the test reflection is introduced at the first point of the threshold curve considered, and the total loudness is calculated. The two stored curves can now be compared in order to calculate the branching angle. Then the next point of the threshold curve is considered. Upon reaching the end, the program turns to the next threshold curve.

In this way all the 45 measured points contained in Fig. 2.10 are treated one by one. For each of these points the program determines how the respective curve of the total loudness branches off from the respective basic loudness curve. The 45 calculated branching angles are listed in Table 4.2. All the measured points of Fig. 2.10 are positioned at fixed spots of the time axis which are numbered as they are sequenced in time.

Of course, Table 4.2 disproves the hypothesis that the drift thresholds are determined by *strictly* constant branching angles, and we have to make a moderation. The hypothesis shall be regarded as confirmed if the straggling of the angles does not exceed 20% of their average 0.085. This corresponds to an angular range from 0.068 to 0.102 (about $5° \pm 1°$). The moderated condition

Table 4.2. Branching angles of time functions of the loudness, in rad

Time axis, spot	1	2	3	4	5	6	7	8	9	10
Reflection 1	.081	.078	.073	.082	.080	.127	.055	.076	.056	.055
Reflection 2		.074	.067	.069	.066	.116	.113	.100	.076	.057
Reflection 3			.122	.101	.098	.095	.080	.096	.060	.077
Reflection 4				.100	.097	.090	.105	.145	.123	.073
Reflection 5					.065	.085	.066	.106	.169	.054
Reflection 6						.066	.053	.051	.102	.058

Table 4.3. Required disadjustment of the level of the test reflection, in dB

Time axis, spot	1	2	3	4	5	6	7	8	9	10
Reflection 1						−1.3	+0.7		+1.1	+1.6
Reflection 2			+0.1		+0.2	−0.9	−0.6			+1.2
Reflection 3			−1.2						+0.7	
Reflection 4							−0.2	−1.9	−0.9	
Reflection 5					+0.3		+0.2	−0.2	−2.1	+1.3
Reflection 6						+0.1	+1.5	+1.6		1.0

is immediately fulfilled by 22 out of 45 angles. In the other cases the level of the test reflection is varied in the computer until the angle falls into the permitted range (see Table 4.3).

The drift thresholds measured in listening experiments are once more shown by the full lines in Fig. 4.4, together with their accuracy range. In order to reach the permitted angular range of the branching theory, the black points in the figure have to be partly shifted. These necessary level shifts of the test reflection are shown by the dotted lines. In most cases the dotted lines remain in the accuracy range of the listening experiment. This confirms the branching hypothesis: *The drift thresholds are determined by the angle between the time functions of the loudnesses*. The required level shift can even smooth out a few conspicuous peaks, as in the lowest two curves. This may create some spontaneous confidence in the theory, without reducing the general trust in subjective measurements.

Each of the full lines in Fig. 4.4 was measured with a variable test reflection R_N, while all preceding reflections $R_{N-1}, \ldots, R_1$ were fixed at the beginnings of the preceding curves (thick points). By this fixing the law of the first wave front could be carefully checked with reasonable experimental effort. However, *a drift threshold is defined for any arbitrary set of preceding reflections*, which may be fixed at their respective drift threshold or at lower level values.

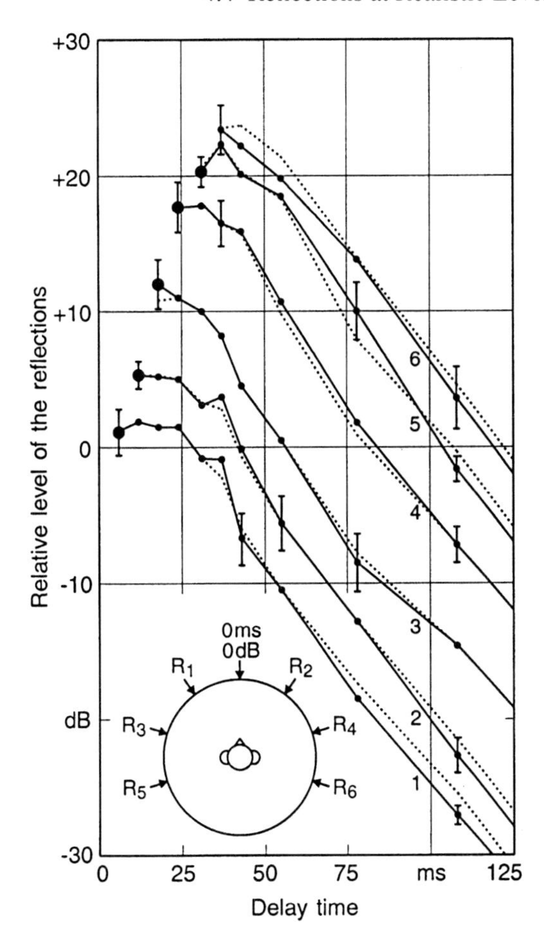

Fig. 4.4. Drift thresholds: — subjective measurement, · · · calculated with the branching theory

In order to dispense with subjective measurements, we can use the computer program to simply calculate the thresholds with the branching theory of time functions of the loudness. Figure 4.4 shows, for a wide range of delay times and level values, that the computed results are as reliable as listening experiments.

4.4 Reflections at Realistic Level Values

The interrelation between computed drift thresholds and computed time functions of the loudness shall now be explained with Figs. 4.5 and 4.6. Taking up the physical conditions of a real room, we consider the direct sound and two

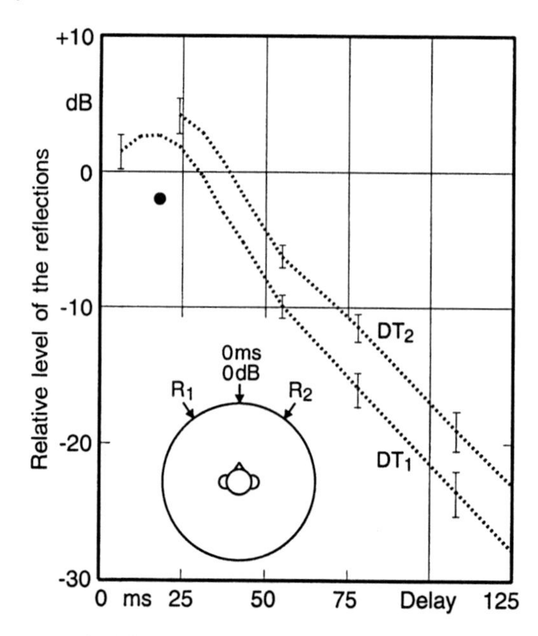

Fig. 4.5. Computed drift thresholds for R_1 and R_2 (DT_2 with R_1 fixed at •)

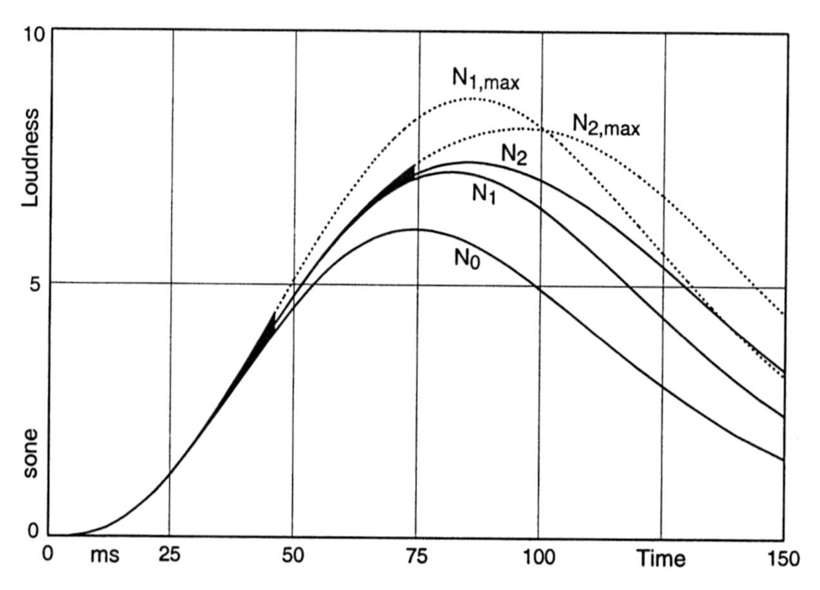

Fig. 4.6. Loudness of the vowel "u" (but). N_0: direct sound 0 ms/68 dB abs.; $N_{1,max}$: with R_1 at 18 ms/+2.6 dB (DT_1); N_1: level of R_1 reduced to −2.0 dB; $N_{2,max}$: plus R_2 at 43 ms/−1.7 dB (DT_2); N_2: level of R_2 reduced to −7.5 dB

early reflections R_1 and R_2, adjusted either to their respective drift threshold or to reduced, realistic level values. The absolute sound pressure level of the direct sound, related to the listener's seat, is assumed to be 68 dB. The time function of the loudness of the direct sound is shown by the basic curve N_0 in Fig. 4.6, once again calculated with the short vowel "u", spoken as in the word "but".

The lower curve DT_1 in Fig. 4.5 shows the computed drift threshold of the first reflection R_1. If its delay is chosen to be 18 ms, the drift threshold is reached at +2.6 dB. The upper curve in Fig. 4.6, printed as a dotted line, shows the total loudness $N_{1,max}$ including the reflection R_1 at this extreme level value. However, in a real room a first reflection at 18 ms delay will certainly be less powerful. Therefore a level value of −1.5 dB was chosen for R_1, indicated by a black spot in Fig. 4.5. In this case, with the realistic reflection R_1, the total loudness N_1 is given by the full line leaping out of the first black wedge in Fig. 4.6. This wedge indicates a branching angle of 5° between the basic loudness curve and the total loudness with R_1 at the drift threshold.

The discussion can now be continued with R_2. For the sake of a clear separation in time the delay of R_2 was chosen to be 43 ms, and the drift threshold of R_2 was computed with R_1 fixed at 18 ms/−2 dB. The result is shown by the upper curve DT_2 in Fig. 4.5. Thus R_2, with 43 ms delay, would reach its drift threshold at a level of −1.7 dB. With the use of this value the dotted curve $N_{2,max}$ shown in Fig. 4.6 was computed. However, in a real room a reflection at 43 ms delay cannot have a relative sound pressure level of −1.7 dB. Thus a more realistic level value of −7.5 dB was chosen for R_2 when computing the curve N_2 shown in Fig. 4.6. This curve includes the direct sound and the two realistic reflections.

Let us now consider Fig. 4.6 more generally, but emphasising the situation near the two black wedges. The area between the curves N_0 and $N_{1,max}$ is subdivided by the curve N_1 leaping out of the first wedge, and the two areas below and above N_1 are of nearly equal size. This confirms the known fact that a first reflection at 18 ms delay—and at a realistic sound pressure level—will have a considerable acoustic effect. The area between N_1 and $N_{2,max}$ is subdivided by the curve N_2 leaping out of the second black wedge, but this is a subdivision at unequal parts. The area between N_2 and N_1 is definitely smaller than the area between N_2 and $N_{2,max}$. A second reflection at 43 ms delay is simply too late for achieving an important effect. This general agreement with practical experience may recommend the time functions of the loudness as a useful tool in acoustical investigations.

Each of the black wedges may be regarded as a source for a set of possible curves and acoustic effects related with the curves. In considering this situation a short digression into Greek mythology may be permitted. Such a wedge reminds us of a "horn of plenty", but the question is: *plenty of what?* Might it be only loudness?

5

Loudness and Diffuseness

5.1 Introduction

The extremely powerful onset of reverberation described in Chap. 2 can certainly stimulate fundamental considerations. This was proved in Chap. 4 with the formal interpretation of the measured drift thresholds, described by constant branching angles between time functions of the loudness. However, we still must examine the background of this interpretation and the subjective sound effects caused by the reflections. Therefore the astonishing drift thresholds shown in Fig. 2.10 and Fig. 4.4 (page 89) shall be used once again for motivation (and they really deserve being considered a second time).

The first sections of the measured curves shown in these figures were determined with a direct sound and one to six early reflections from different directions. In this respect the diagram complies with the onset of reverberation in concert halls. In the experiments, however, all reflections had extremely high levels adjusted to their respective drift threshold. The first points of the threshold curves especially represent an impulse response with a total duration of only 37 ms, beginning with the direct sound. This fast sequence of pulses with higher and higher energy reminds us more of a thunderstorm than of the sounds in a concert hall. The first point of the highest measured drift threshold, for example, concerns the sixth reflection at a delay of 37 ms and a relative sound pressure level of +23.4 dB. If we try to relate this measured point with an arbitrary seat in an arbitrary concert hall we must admit immediately that this is impossible. In a concert hall (without vaulted sound mirrors) the natural sound propagation would generate a reflection level near −4 dB. Because of the large discrepancy of more than 27 dB, the practical relevance of the experiments described in Chap. 2 must be doubted.

In the search for an extensive explanation of the drift thresholds, apart from the formal interpretation with the branching theory, just the large discrepancy achieved with six reflections might be enticing. Basically, however, the problem of a proper explanation will arise with only the first sound reflection if it is adjusted to the drift threshold. Therefore, avoiding a branched discussion, we shall continue with this simple case.

5.2 Amounts of Neural Excitation

Let us consider a sound field that consists of the direct sound and only one reflection. For this case, two time functions of loudnesses are depicted in Fig. 5.1, printed as full lines. The curves are copied from Fig. 4.6, and as mentioned earlier they were computed with the impulse-like vowel "u", spoken as in the word "but". The basic curve shows the loudness of the direct sound at an absolute sound pressure level of 68 dB, related to the listener's place. The upper curve represents the total loudness including a reflection at a delay of 18 ms and a sound pressure level of +2.6 dB, related to the direct sound. This is the level of the drift threshold DT_1 at the delay time chosen (Fig. 4.5). Thus the drift threshold is not exceeded, and the perceived posi-

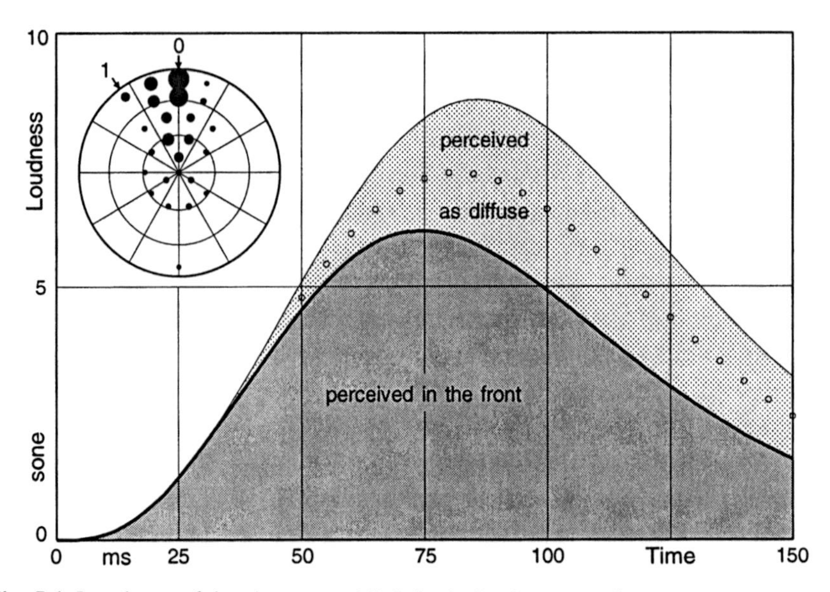

Fig. 5.1. Loudness of the short vowel "u" (but). *Basic curve*: direct sound. *Top curve*: with reflection 18 ms at the drift threshold; o o o reflection at realistic level −2.0 dB. Subjective sound distribution: with reflection 6 ms at the drift threshold (Fig. 3.6.1).

tion of the talker remains at the azimuthal angle 0°. This is on the subject's line of view.

Considering the basic curve shown in Fig. 5.1 the reader may imagine a huge number of neurons in the auditory system, permanently reporting "frontal sound". To describe this activity, A. Vogel and E. Zwicker introduced the term "neural excitation" [6, 7]. The term may be applied directly to the dark grey area below the basic curve of Fig. 5.1. The light grey area, too, means many active neurons, but they *cannot announce "frontal" sound*, because the difference between the two curves is caused by the reflection arriving from another direction. However, the neurons generating the light grey area *cannot announce a new main direction* in the sound image either. The reason is that the drift threshold was not surpassed, as the upper curve in Fig. 5.1 was calculated with the reflection *at the drift threshold.*

This seeming contradiction disappears when considering the subjective sound image determined by a group of 21 subjects (circular diagram). Obviously, the light grey area between the two curves indicates amounts of neural excitation causing an *enlargement* of the direct-sound image just arising. This takes place more or less on all sides, in an indefinite, diffuse way. The use of different delay values in the figure is not decisive.

In order to summarize we can say *that the dark grey area concerns the talker position perceived in the sound image, whereas the light grey area indicates the diffuseness of the image.*

5.3 Loudness Diagram versus Impulse Response

Placing a dummy head with implanted microphones (see Chap. 1) at a certain seat of a concert hall, the impulse response of the room can be binaurally recorded. To make the recording we excited the room with a pistol shot. This recording contains *the full acoustical information* related to the selected seat. However, the problem of interpreting the large number of needlelike lines in the impulse response, shown as a diagram, may be more difficult than the direct judging of the original room by ear.

In principle, impulse responses are abstract mathematical functions, defined with a "δ-pulse" of duration zero. When trying to approximate this case experimentally, the lines in the impulse response become sharper and incredibly numerous, but the redundancy in these many details becomes more and more obvious. Thus any evaluation of the lines must include a reduction of redundancy.

When a working hypothesis is suggested for this purpose, it should be focused on the fact that the decisive acoustical judgement must be left to the listener's ear. Concisely, when analyzing an impulse response, ignoring things which cannot be subjectively perceived may be recommended. In other words, the directly noticeable quantities have to be preferred; one of these is the loudness, to which we now return.

Modern computers can, of course, handle any number of complicated short impulses and sound signals. However, a computer program calculating time functions of the loudness must include the integrating effect of the ear. Thus, the computer evens out and summarizes the details of the impulse response, and the result is a smooth curve. This does not mean denying the importance of short impulses in hearing, especially the spectral aspect. As mentioned earlier, continuous speech or music have immensely varying time functions of the loudness. In the course of these smooth functions, the ear searches for sections with high positive curvature. They are caused by strong, impulse-like sounds in the signal, and these parts of the curves control directional analysis (Chap. 2).

5.4 Discussing Neural Excitation

5.4.1 Effect of a Single Reflection

Using the described computer program for the loudness, any given impulse response of a room may be transformed to a diagram like Fig. 5.1. This representation clears away some redundant details of the impulse response, and the essential information becomes more evident. Therefore Fig. 5.1 may serve as a suitable base for further considerations. The diagram offers the possibility of comparing the two grey areas delimited by the curves. As explained earlier, these areas indicate the amounts of neural excitation which generate a frontal sound impression on the one hand, and the diffuse parts of the image on the other hand. Acting on a trial basis, we estimate the sizes of the areas, but only between 0 ms and a line at 90 ms which is defined by the point of maximum total loudness. Thus we calculate the "diffuseness quotient" $D_N = A_{\text{light grey}}/A_{\text{total}} \approx 0.16$.

This result recalls the diffuseness value $D = 0.09$ from Chap. 3, determined with the circular diagram shown in Fig. 5.1. That diagram, too, concerns a sound field with a single reflection, its level being at the drift threshold. Though the delay values for the two parts of Fig. 5.1 do not quite agree, this curious comparison may emphasize the interconnection of Chaps. 2 to 4.

The diffuseness D as defined in Chap. 3 is strictly based on listening results, whereas the diffuseness quotient D_N derived from time functions of the loudness (neural excitations) remains abstract.

However, the time functions of the loudness plotted in Fig. 5.1 indicate the dynamic aspect of diffuseness. When the pulse has reached the ear, the basic loudness rises; but up to a time of 45 ms the diffuseness remains "zero" (in accordance with Sect. 3.3 this means $\approx 5\%$). Only after the beginning of the branching of the curves does the sound image widen while the loudness increases further. Maximum diffuseness, in this case, is reached at about 90 ms—the moment of maximum total loudness. While the loudness fades out, the diffuseness does not change in the main. These processes are too fast to be observed step by step.

As explained earlier, the diffuseness quotient $D_N = 0.16$ was determined with a single reflection at a delay of 24 ms and adjusted to the drift threshold. Figure 5.2 shows, in accordance with common experience, that the reflection generates no diffuseness if its delay is more than 50 ms.

Thus a reflection *can or cannot* generate diffuseness, even though it is always maintained at the drift threshold (see, e.g., Fig. 2.4, page 52). This fact is certainly astonishing for the reader. However, the seeming contradiction disappears when carefully regarding the criterion used for the measurement of the drift thresholds (Chap. 2). The decisive question was: "Does the direction of the greatest loudness agree with the line of view, and is there no echo to be noticed?" As the drift thresholds were measured by the author himself we can be sure that the linguistic distinction in the sentence was always regarded.

Beginning at about 50 ms delay, in a sliding transition, the second part of the question becomes more and more important. The risk of perceiving

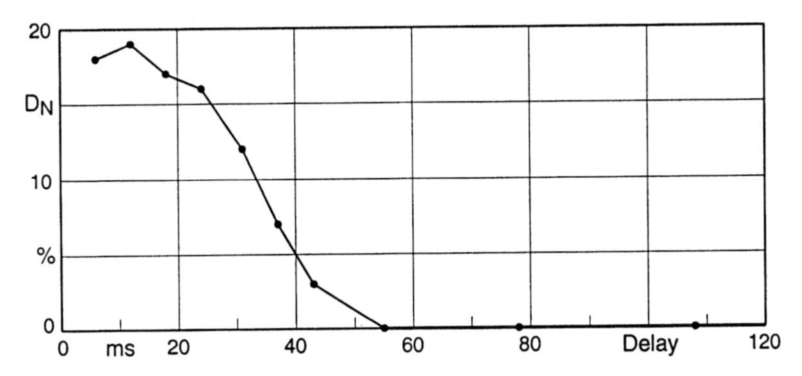

Fig. 5.2. Diffuseness generated by a single reflection (drift threshold); D_N is derived from time functions of the loudness (neural excitation).

an echo is very high, indeed, if a separate reflection is delayed by more than 100 ms, and this would include a directional jump in most cases. Fortunately, however, *the theory of branching angles between the time functions of the loudness works likewise with both parts of the question*; in short: Fig. 4.3 $\Rightarrow$ Fig. 4.4. The discussion of this problem with more and more reflections would be very complicated and may be omitted.

We shall now try to find another approach to the practical application of Fig. 5.1. Let us consider a listener in a large concert hall, seated at a distance of 24 m from the stage. In relation to the direct sound, the first reflection may arrive with a delay of 18 ms, the same as used for the figure.

The mentioned delay value corresponds to an additional sound path of about 6 m. Thus, the relative sound pressure level of the reflection will be $24 \text{ m}/(24 \text{ m} + 6 \text{ m}) = 0.8 \approx -2$ dB, caused by the natural sound propagation. This value was used when computing the line of circles in Fig. 5.1. The remaining light grey area below this line agrees with the experience that, even under realistic conditions, the onset of reverberation shows a certain degree of diffuseness.

5.4.2 Six Strong Early Reflections

In a quick, introductory explanation, the decay of the sound energy in the genuine reverberation of a room may be described with the exponential function $e^{-t/\tau}$. The onset of reverberation is the mathematical opposite, and therefore it might be described with the term $1 - e^{-t/\tau}$. However, a more realistic description of these processes must start from the fact that the sound energy is subdivided into many portions, called reflections, because the sound is reflected from the walls or the ceiling. In the decaying, genuine reverberation, when the sound is stochastically mixed after many reflections, this discontinuity is of minor importance.

On the other hand it is particularly interesting to consider *the rise of the loudness in the onset of reverberation* (see Fig. 5.3). The process was calculated with six powerful reflections, rapidly following each other in 6 ms steps; they were introduced in Chap. 2 and discussed earlier. The subsequent genuine reverberation has no part in this process, because the impulse response ends after 37 ms.

Once again the impulse-like "u" (as in "but") was used for the calculations. All levels of the reflections were adjusted to their respective drift threshold. Thus, all branching angles between the curves are $5° \pm 1°$.

The loudness rises steeply, and a high total loudness is built up. The small circular diagram at the top left shows that the reflections generate high dif-

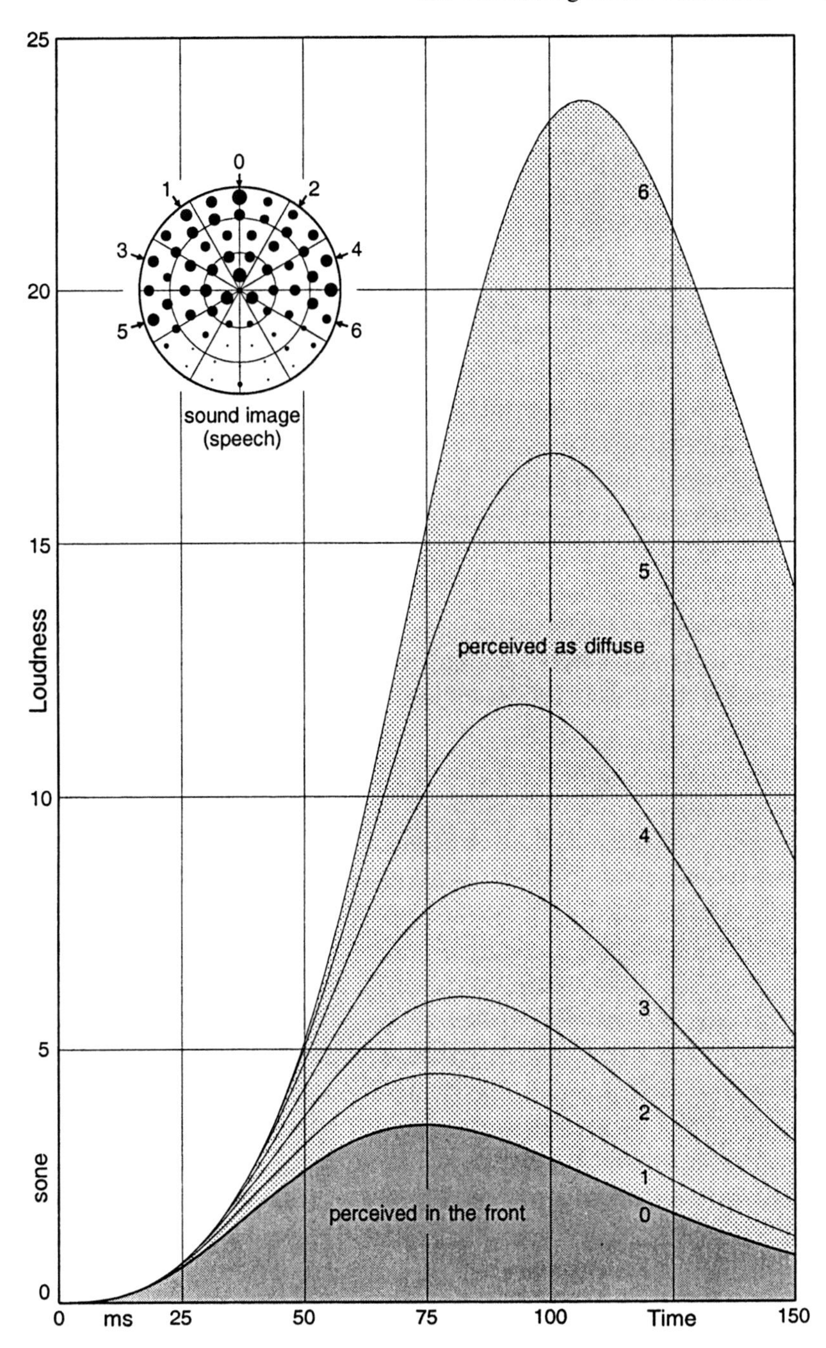

Fig. 5.3. Loudness of the short vowel "u" (but), direct sound and six reflections. All reflection levels at their respective drift threshold. ⇒ All branching angles 5° ± 1°.

fuseness with continuous speech. In this case the interesting diffuseness quotient $D_N = A_{\text{light grey}}/A_{\text{total}} \approx 0.84$ is not far from the value $D = 73\%$ mentioned in Chap. 3. The degree to which each single reflection contributes to the diffuseness may be estimated by considering the regions between the curves.

The pile of loudness curves resembles a group of acrobats in a circus, forming a pyramid of bodies by climbing onto each others' shoulders. The sound image of this excessively powerful onset of reverberation is shaped like a bowl over the head. The size of the bowl roughly agrees with the listener's distance from the loudspeakers. The thickness of this layer of sound cannot be determined. It would be inadequate to designate that sound impression as spacious; the word diffuse is more suitable.

As discernible in Fig. 5.3, the total loudness reaches a maximum of 24 sone. This corresponds to a sound pressure level of about 86 dB; thus the speech is "very loud". If the six zigzagging reflections are switched off, all at once, the frontal loudspeaker seems to whisper. However, if the lateral reflections are added again, partly being extremely strong, the frontal component plays a normal part in the loud sound, and a directional equilibrium is maintained in the subjective impression.

5.4.3 Boston Symphony Hall

The impulse response of a narrow and high cuboidic room, shaped like Boston Symphony Hall, initiated the discussion of Chaps. 4 and 5. Now, towards the end of this discussion, it is certainly interesting to consider time functions of the loudness, calculated with the data of this famous hall [33], and to know the diffuseness quotient D_N derived from those curves. This time, however, the room is regarded at original size.

The curves of Fig. 5.4 were calculated with the computer program presented in Chap. 4. In order to simplify the explanation of the diagram we may compare it with Fig. 5.1; in both cases an absolute sound pressure level of 68 dB was used for the direct sound.

The single reflection considered in Fig. 5.1 had a delay of 24 ms and was adjusted to the drift threshold. In Boston Symphony Hall, however, the first reflection shows values of 22 ms/−2.2 dB, which is more than 4 dB

Table 5.1. Early reflections in Boston Symphony Hall

Delay time:	22 ms	38 ms	47 ms	66 ms	76 ms	85 ms
s.p. level:	−2.2 dB	−3.6 dB	−8.0 dB	−7.5 dB	−9.5 dB	−11.5 dB

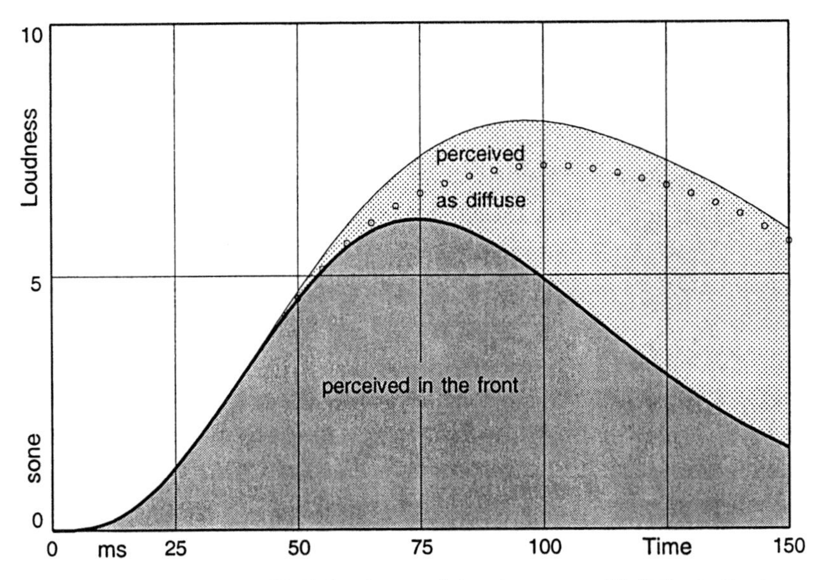

Fig. 5.4. Boston Symphony Hall, loudness of the short vowel "u" (but). *Basic curve*: direct sound. *Top curve*: with reflections 1 to 6; o o o reflection one off.

below the drift threshold. Therefore the maximum total loudness is lower in Fig. 5.4, and it is reached 6 ms later, at 96 ms. Furthermore, caused by the next reflections, the total loudness fades out rather slow in Boston Symphony Hall. As the diagram includes three reflections at 66/76/85 ms the transition to genuine reverberation is already discernible.

The diffuseness quotient for Boston Symphony Hall, determined within the limits 0 and 96 ms, is $D_N = A_{\text{light grey}}/A_{\text{total}} = 14.7\%$. If the first reflection at 22 ms is switched off (line of circles) the maximum total loudness is lower, and it is reached only at 100 ms. In this case the diffuseness quotient, determined between 0 and 100 ms, decreases to $D_N = 10.4\%$. These results appear plausible, and thus the theory of loudness and diffuseness has passed a test of practical application.

5.5 Interaural Cross-correlation

The acoustical quality of a concert hall decisively depends on the existence of powerful early reflections, and the degree of their influence has been extensively discussed in scientific publications. Some well-known examples are: a numerical definition of clarity [4], measurement of speech intelligibility [41, 42], a quality scale based on general experience [43], factor analysis based on comparison of music recordings [10]. A survey on this research may be looked up in textbooks on acoustics [44]. All these publications agree in the conclusion that the early reflections are of fundamental importance.

The investigations presented in this book concerning drift thresholds, time functions of the loudness, and the diffuseness of sound images may be categorized in this scientific tradition. However, the investigations described here are straightly aimed at processes in the listener's ear.

In the human auditory system, the signals received at the ears are analyzed with basically three different physical "mechanisms" [45–47]. The first one, realized on the basilar membrane in the cochlea, and in the nuclei immediately following, performs spectral analysis. As explained in Chap. 4, it can be compared with a set of self-adapting $1/3$ octave filters. The second mechanism, installed in the lateral lemniscus assigned to each ear, carries out running autocorrelation with the signal received at the respective ear. The result of this process is needed for pitch perception, which may be regarded as a completion of spectral analysis. Third, the two signals received at the ears are compared in the inferior colliculus next to the auditory cortex. This large arrangement of neurons carries out the process of interaural cross-correlation.

The decisive part of its mathematical formulation is

$$\Phi(\tau) \sim \int f(t)g(\tau - t)\,\mathrm{d}t,$$

where $f(t)$, $g(t)$ are the time functions of the signals received at the ears, and τ denotes a time shift; details are explained in the following.

5.5.1 Cross-correlator as a Neural Device

The idea that autocorrelation might have a part in human hearing was presumably introduced by J.C.R. Licklider in 1951 [45]. He included cross-correlation a few years later [46]. An auditory cross-correlator, realized by chains of neurons, may be imagined as consisting of two antiparallel bucket brigades (Fig. 5.5), each of them provided for one of the signals $f(t)$, $g(t)$. The signals enter the bucket brigades from opposite sides, and thus no time inversion is required for the minus sign in the expression $g(\tau - t)$ of the correlation integral. Let us imagine each bucket fixed at its place, and the contents

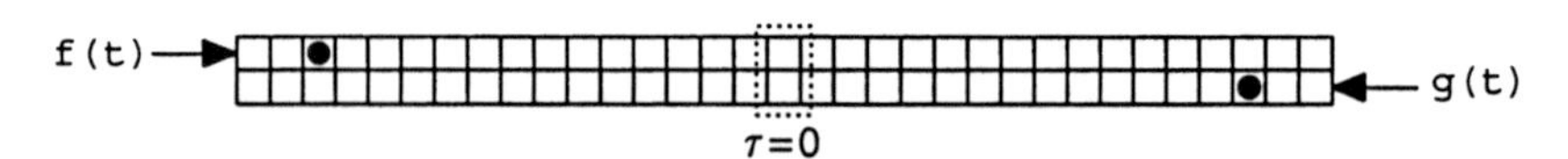

Fig. 5.5. A sound impulse received from the front, travelling in two antiparallel bucket brigades of an auditory cross-correlator.

being poured over from bucket to bucket, depending on a constant clock frequency. If each chain has 33 buckets, and if the signals are sampled every 50 μs, they need 1600 μs for completely passing the chain; the last bucket is simply emptied.

When a short sound impulse reaches the listener's head from the front, two equal pulses begin to travel in the antiparallel chains and meet each other at the central pair of buckets. This spot defines the time shift $\tau = 0$. The τ-axis extends from -800 μs to $+800$ μs, covering all possible values of the interaural time delay for sound incident from whatever direction around the listener's head. The two hemispheres correspond to the sections of the cross-correlator and can be distinguished by the sign of τ.

At the spot $\tau = 0$, *and also at any other spot of the double chain*, the two passing signals are multiplied, and the result is subject to running short-time integration. In this process the product $f(t)g(\tau - t)$ occurring at the present has the greatest weight, and the more a product lies in the past, the more it is forgotten in the integral. Mathematically, this feature is achieved by a decaying exponential weight function. In the example of Fig. 5.5 all these integrals are zero, except for the pair of buckets at $\tau = 0$.

Concerning the feasibility of the correlator, a few remarks on neural data processing might be helpful. The signal delay required in the process is given by the travelling time in the dendrites, when neural peaks are passed from neuron to neuron in a neural chain. Furthermore, by increasing their activity, neurons are capable of adding up several peaks arriving from other neurons; this capability is sufficient for multiplying signals, too. Finally, the activity of a neuron fades out automatically if no further information is received. Thus, all mathematical elements required for running short-time correlation are given.

5.5.2 Measured Cross-correlation, One Sound Direction

In the following discussion about loudness and diffuseness, interaural cross-correlation is the centre of attention because it allows left–right discrimination in directional analysis and thus concerns the diffuseness of sound images. Front–back discrimination is based on pitch perception and spectral analysis. These aspects are also relevant in directional analysis, but they shall be disregarded in the following.

The curve of Fig. 5.6 was measured with using a dummy head in an anechoic chamber [28], and it is valid for wide band noise limited at 0.2 kHz and 6.3 kHz. The dummy head received the sound at an elevation angle of 0°. The main maximum of the curve is positioned at a certain value of τ from

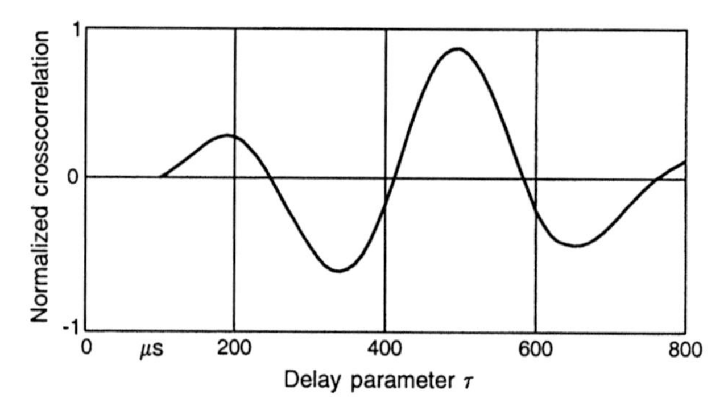

Fig. 5.6. A measured interaural cross-correlation function.

which the ear can derive the azimuthal angle of the sound. At large angles of incidence the diffraction at the head causes both an interaural time delay and a marked linear distortion of the signal; thus, the main maximum is <1.

5.5.3 Calculated Cross-correlation, Three Sound Directions

We investigate the relation between diffuseness and interaural cross-correlation by comparing two special sound fields. Each field is constructed with a direct sound and two early reflections, arranged differently in the two cases (see Fig. 2.12, page 63).

However, in order to allow a clear discussion, a few drastic simplifications must be proposed concerning cross-correlation diagrams like Fig. 5.6. First, the height of the main maximum, being not much smaller than 1 for most sound directions, proves that the signals received at the two ears do not differ to a relevant degree, apart from their mutual delay. Therefore we can replace

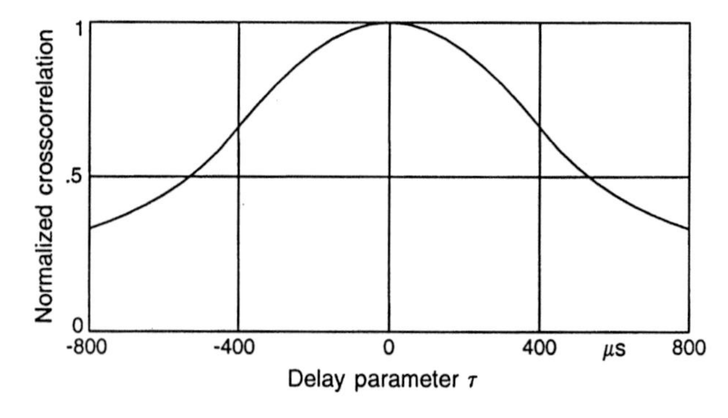

Fig. 5.7. Simplified interaural cross-correlation function, bandwidth 1200 Hz.

the cross-correlation function by the average of these signals' autocorrelation functions. If the main maximum is placed as before, the result is a symmetric curve resembling to a large extent the cross-correlation function. Second, only the absolute values of the function are considered, because negative neural excitation is impossible. Finally, the detailed rise and fall of the function shall be replaced by its completely smooth envelope [47].

Figure 5.7 shows the effect of these simplifications. The example was computed with a noise signal which has a flat spectrum, as in Fig. 5.6. However, the bandwidth was reduced to only 1200 Hz, and thus the signal is assumed to represent the short vowel "u" of the word "but", spoken with an accentuated articulation [34]. As mentioned earlier, this sound is suitable for replacing continuous speech in loudness calculations.

If the listener's head is exposed to a direct sound and two early reflections, each of these components develops its own neural activity in the interaural cross-correlator. The activities are concentrated at different spots of the antiparallel bucket brigades (Fig. 5.5), and these spots indicate the azimuthal angle φ of the respective sound field component. With the elevation angle zero, the interaural time delay τ and the azimuthal angle φ are related by $\tau \approx 255\ \mu\text{s} \cdot (\varphi + \sin\varphi)$, depending on the size of the head.

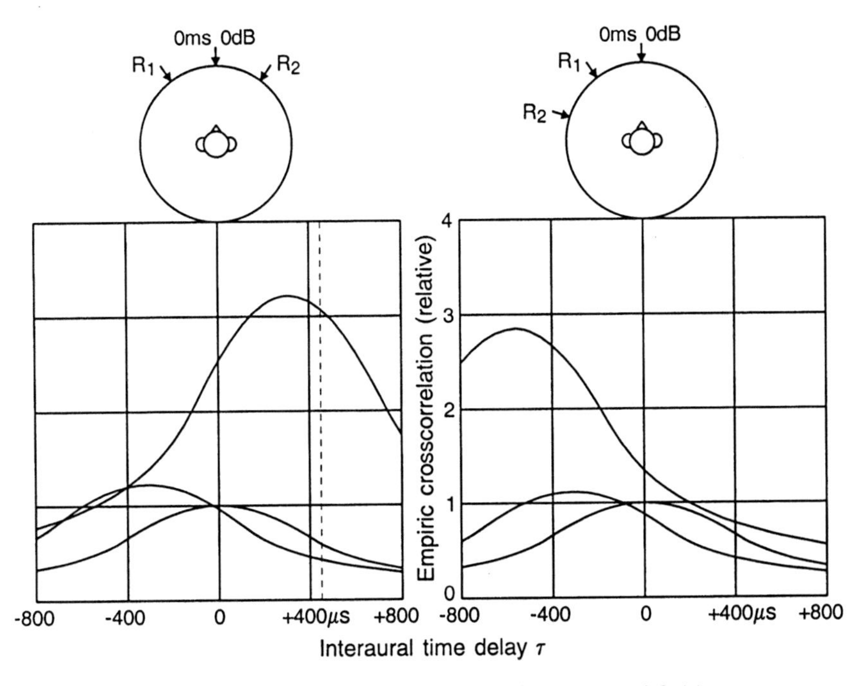

Fig. 5.8. Interaural cross-correlation functions of three sound field components.

In both parts of Fig. 5.8 the direct sound (0 ms/0 dB) generates a function centred at $\tau = 0$, and the reflection R_1 at $\varphi = +36°$ generates a maximum at $\tau = -310$ µs. These two maxima are nearly equal in accordance with the relative sound pressure levels +1.1 dB and +0.7 dB of R_1.

The decisive point in the diagrams is the contrast in the arrangement of R_2. Its azimuthal angle was −36° and +72° in the two cases, and the two highest maxima are positioned accordingly. The respective sound pressure levels of R_2 are +5.3 dB and +5.0 dB. (The maxima depend a little on the azimuthal angle [28]: −5% at ±36°, and −10% at ±72°.)

With respect to the direct sound the reflections R_1 and R_2 are delayed by 6 ms and 12 ms, and the neural activities in the bucket brigades must develop with quite the same time lags. However, the vowel "u" (spoken as in the word "but") lasts about 50 ms, and thus its cross-correlation functions overlap at least partly in time.

5.5.4 Neural Cross-correlator as a Loudness Analyzer

Concerning the unit of measurement, Fig. 5.8 brings about the following considerations. The maximum absolute sound pressure level to be processed by the ear is about 160 dB (jet engine), where 0 dB means the lower limit of hearing [40]. The value 160 dB corresponds to a numerical ratio of 10^8 to 1 for the signal amplitudes, and for the energies we have even 10^{16} to 1. Correlation functions, if not normalized, actually have the dimension of an energy. This results in some problems which may be explained with Fig. 5.8. If we agree that 0 dB means 1 mm in the diagram, the sound of a jet engine would be represented by 10^{16} mm $= 10^{10}$ km! This incredibly high numerical ratio leads to the conjecture that a neural cross-correlator can exist only in a modified form. In loudness analysis (simplified according to ISO standard, see the computer program in Chap. 4) the sound pressure amplitudes are raised to the power of 0.602, which means 0.301 for the energies. With the use of this standard exponent the aforementioned ratio decreases to "only" 65000 to 1, *provided that neural cross-correlation and loudness analysis are merged to one and the same process.*

The definition of correlation functions includes an integration over the time extending from the present to infinity. In neural data processing, however, the infinite integration is reduced to a short time interval, and the mathematical definition of correlation must be adequately modified. This transition from infinite integration to running short-time processes was described by B.S. Atal, H. Kuttruff, and M.R. Schroeder [48].

The scheme of a neural cross-correlator presented in Fig. 5.5 contains two antiparallel bucket brigades and a corresponding set of integrators, each of them provided for one pair of buckets. In an electronic model of this device, proposed by B.S. Atal, the running short-time integration is realized with simple RC elements (resistor and capacitor). Their impulse response is a decaying exponential function, and thus they can store and forget information, just as demanded in Sect. 5.5.1.

At this point of the discussion the reader will certainly remember the simple loudness model proposed by H. Niese (Fig. 4.1, page 81). This model is also based on RC integration, and if the time constants of the neural integrators at the bucket brigades are in the range 28–35 ms they would exactly match the proposal of Niese, but let us go one step further. The calculation of time functions of the loudness described in Chap. 4 might be regarded more or less as a detailed completion of the RC integration, and thus we can summarize.

As concerns the realization in a neural system, there is no obvious reason why loudness analysis should not be merged with interaural cross-correlation; loudnesses cannot be properly determined on a mono base! A modified cross-correlator might thus continuously determine the varying directions of all sound field components and also their partial loudnesses which have to be finally combined to the general loudness.

This structure reminds us of the *spectral* decomposition of sound signals in the loudness analyzer described by Zwicker and Vogel [6, 7]. In that model, too, the general loudness is determined by combining partial loudnesses (from different frequency bands).

Note that the discussion of cross-correlation in directional analysis involves a restriction to left–right discrimination, and that the discussion of front–back discrimination was purposely excluded in this book.

5.5.5 Testing the Generalized Interaural Cross-correlator

A test of the generalized cross-correlator—including a loudness analyzer but disregarding spectral decomposition—leads us back to the correlation functions shown in Fig. 5.8. In each part of the figure the positions of the three maxima roughly agree with the arrows and the letters at the respective circular diagram. These diagrams show the directional arrangement of the direct sound (0 ms/0 dB) and the reflections R_1, R_2 in the two sound fields considered.

However, it is sufficient to discuss mainly the left part of Fig. 5.8. We imaginge the τ-axis subdivided into 16 sections of equal width and select the section marked by the dashed line. The points at which the curves intersect this line indicate three unnormalized correlation values. They are assigned to the respective sound field component and show the energy portions falling into the marked section.

With the use of the exponent 0.301 just mentioned, the three energy values are very roughly converted into loudnesses that must be finally recombined to an overall loudness. However, this result remains limited to the considered section of the cross-correlator. In the recombination the direct sound and the reflections are represented with different weight factors. Their determination can be followed up with Fig. 5.3 (page 99).

After about 80 ms the direct sound has reached its maximum loudness of 3.5 sone. If the reflection R_1 is added, the maximum loudness increases by 1.0 sone, and R_2 causes another increase of 1.5 sone. These values reflect the fact that any delay of a reflection involves a reduction of its effect. Therefore the weight factors 7, 2, 3 are assumed to be roughly suitable for the recombination of the partial loudnesses in the interaural cross-correlator, and they were used

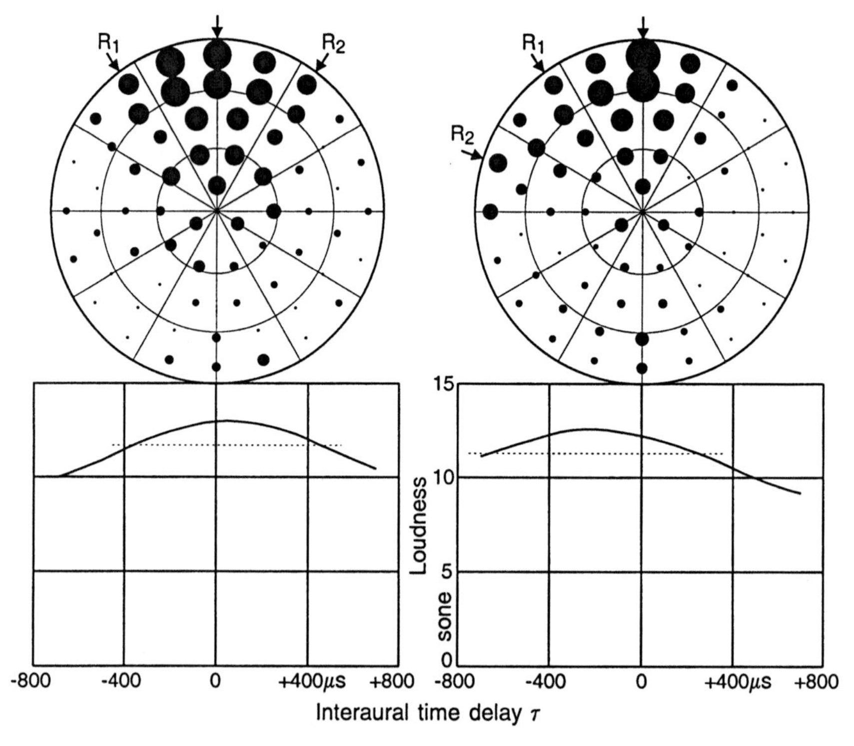

Fig. 5.9. Subjective sound distribution versus loudness distribution in a generalized interaural cross-correlator/loudness analyzer (hypothetical).

in all its sections. Figure 5.9 shows two calculated loudness distributions and the corresponding diffuse sound distributions, acquired with groups of 20 and 21 subjects.

The subjective sound distributions are a copy of Fig. 2.12 (page 63), and the experimental data are specified in that figure. The same data were used for the calculation of the weight factors, except for irrelevant level differences ($\leq$0.4 dB) in the right part of Fig. 5.9. The numeric relations between the factors are maintained if the absolute sound pressure level of the direct sound is changed from 60 dB (Fig. 5.3) to 68 dB (Fig. 5.9).

The calculated loudness distribution in the left part of Fig. 5.9 shows a maximum near $\tau = 0$. This agrees fairly well with the corresponding subjective sound distribution. However, for full agreement the maximum should be shifted slightly to the left. The dotted line indicates equal loudness at $\tau = -360$ µs and at $\tau = +456$ µs. These interaural time delays correspond to azimuthal angles of 42° on the left side of the head, and 55° on the right. Compare this to the subjective sound distribution!

The dotted line in the right part of Fig. 5.9 indicates equal loudness at $\tau = -660$ µs and at $\tau = +253$ µs, corresponding to azimuthal angles of 90° on the left side of the head and 29° on the right. The subjective sound distribution roughly confirms these angles, but the maximum of the right curve should be shifted slightly to the right.

These discrepancies remind us of the still-unexplained push effect mentioned in the description of Fig. 2.11 (page 59). The effect concerns the "centre of gravity" of a subjective sound distribution: the latest and strongest reflection of a sound field tends to push the centre away to the opposite hemisphere. In Fig. 2.11 the effect is obvious in the partial diagrams number 2, 3, 4, 5. In Fig. 5.9 the part of the latest and strongest reflection is taken by R_2, and if the push effect could be introduced into the calculations the maxima of the loudness distributions would slide in just the desired direction!

5.6 Final Remarks

Encouraged by the good agreement between the subjective sound distributions and the calculated curves of Fig. 5.9, we can summarize that loudness analysis and directional analysis might indeed be combined in a modified neural cross-correlator. Thinking about the number of neurons required for the data processing, this solution appears to be rather economic.

However, in the model just described the weight factors are imagined as percentages of the total loudness which is the *output* of the complete analyzer.

On the other hand, these factors are already needed *during* the calculation of partial loudnesses slowly developing in the sections of the cross-correlator. The obvious inconsistency in this procedure might be avoided with an extended version of the computer program for the calculation of time functions of the loudness. The modified program would have to compute the time function of the partial loudness in every section of the cross-correlator, permanently combining these functions to a total loudness function and permanently determining varying weight factors to be used in this continuous process. This scheme is not realized in the computer program of Chap. 4, and thus the required weight factors were simply derived from the final result (Fig. 5.3).

The development of a detailed, extended computer program, modelling the combined cross-correlator/loudness analyzer, is certainly an interesting topic for further investigations. The next step could be a computer model that realizes a spectral as well as a directional decomposition of binaural signals received by a dummy head. The program would have to recombine the components to precise time functions of the loudness for any given sound direction, and also for complex sound fields with many directional components. Furthermore, the program might bring out directional information or "subjective" sound distributions, possibly even including front–back discrimination.

Summary

Directional hearing is based on the geometrical arrangement of the ears on the head and on the strange shape of the outer ears. The head, being an "acoustic obstacle", has been appreciated as an advantage for stereophony since 1970. In taking up this idea, Chap. 1 presents a set of precise devices. The devices consist of a dummy head, an electronic compensation filter, and two loudspeaker boxes having a certain directivity. They are meant to be placed in a bookshelf. With this head-related system, a 360° transmission is realized in a living room. The equipment produces clear improvements for common stereo recordings, too.

The other chapters concern physical processes and subjective sound effects in concert halls, with emphasis on the subjective effects. The findings suggest that they are of general importance.

The reverberation of large rooms appears to be agreeably diffuse. In order to get the same effect for the "onset of reverberation" in concert halls, strong reflected sound components must arrive at the listener's ears immediately after the direct sound. The simulation of such processes in an anechoic chamber is described in Chap. 2. A maximum of six reflections are simulated step by step, their respective levels being increased up to a limit still tenable. These "drift thresholds" are measured, and the more or less diffuse "sound images" are investigated with groups of subjects.

Starting from this point, the "diffuseness D" is defined by a computer program in Chap. 3. Diffuseness values are calculated for the sound images observed at the onset of reverberation. The diffusity d commonly discussed in room acoustics has to be clearly distinguished from the subjective diffuseness D, as serious discrepancies may occur; for example, $d = 6\%$ but $D = 73\%$ in one and the same sound field. The importance of the diffusity d

for the subjective impressions becomes doubtful at such contrary numerical values.

In Chap. 4, the drift thresholds of the reflections are related to the time functions of the loudness. The loudness is calculated with the short vowel “u” as spoken in the word “but”. It turns out that the measured drift thresholds can be calculated with the loudness curves for this vowel.

The fifth and last chapter shows that the diffuseness, too, can be explained with time functions of the loudness. However, interaural crosscorrelation, being a tool in directional analysis, is also involved in the formation of diffuse sound images. These mutual relations are closely considered, and it becomes obvious that cross-correlation and loudness analysis might be directly connected in one and the same neural process.

All five chapters point out the dominant part of the auditory system in acoustics: Sound images are formed only by data processing ***in the listener’s head***, and this “device” always feels like doing conjuring tricks.

Appendix

```
'**********************************************************************
'           BASIC program for time functions of the loudness
'       based on B( ),But( )= short version of the vowel "u"(but)
'**********************************************************************
 CLEAR 50000,50000: CLS: COMPILER "FPU2;68882;CUTLIB";
 DEFSNG "A,B,C,D,E,N,P,V": DIM T(10),Db(10),Amp(10)
 DIM B(160),But(160),N(161),Nzero(161),Nplus(161)
'--------------------------------------------------- delay times in ms
 T1=6:T2=12:T3=18:T4=24:T5=31:T6=37'                fixed reflections
 T(1)=6:T(2)=12:T(3)=18:T(4)=24:T(5)=31'         variable reflections
 T(6)=37:T(7)=43:T(8)=55:T(9)=78:T(10)=108
'---------------------------------------------------------------------
'FOR T=1 TO 160:B(T)=1.: NEXT '           square 160ms for calibration
 FOR I=1 TO 105: READ B(I): NEXT '          shortened "u", uncalibrated
-Begin: Vari=0: Out$=" ": PRINT "Direct sound:     dB"
 INPUT @(0,14);Db0:Db0=Db0+3'                      +3dB spectral extra
  P=10^(Db0/20):Calib=6.125E-3*P'          square 160ms,40dB ==> N=1
  FOR T=1 TO 160:But(T)=Calib*B(T): NEXT'         short "u", calibrated
 INPUT "Reflection:          ";Ref' ------------------------------------
 FOR S=Ref TO 10                        S = delay = spot on the time axis
-Again:Clear5: USING : FOR T=0 TO 160:N(T)=0: NEXT '            steps 1ms
 IF Vari=0 THEN PRINT "Working on basic curve...";
 IF Out$="c" OR Out$="a" THEN PRINT "Working at spot";S;"...";
 IF Out$="s" THEN ' on screen: variable refl.level --> dev.angle
    PRINT @(5,14);"(0=end)";
    INPUT @(5,0);"Spot";S: PRINT @(5,14);"       ";
    IF S>0 THEN S= MAX(Ref,S):S= MIN(S,10) ELSE STOP
    INPUT @(6,0);"Additional level, dB: ";Vari:Clear5
    PRINT "Working at spot";S;", addit.level";Vari;"dB...";
    Vari=10^(Vari/20) ENDIF '              additional level as a factor
'****************** calculating N in steps a,b,c,d,e ******************
'a) ---------------------- total sound pressure = SQR(sum of energies)
 SELECT Ref' = Reflection
 CASE 1' Ref 1 variable, at the spots S=1...10 of the time axis
  'Db( )= drift threshold of Ref 1 according to Fig.2.10
   Db(1)=1.1:Db(2)=1.9:Db(3)=1.5:Db(4)=1.5:Db(5)=-.8' <--Fig.2.10
   Db(6)=-.9:Db(7)=-6.7:Db(8)=-10.5:Db(9)=-18.5:Db(10)=-27.1
   FOR I=1 TO 10:Amp(I)=10^(Db(I)/20): NEXT ' amplitudes
   FOR T=1 TO 160:N(T)=But(T)^2' sum of energies
```

```
    IF T-T(S)>0 THEN N(T)=N(T)+(Vari*Amp(S)*But(T-T(S)))^2
    N(T)= SQR(N(T)): NEXT ' here N(T) means sound pressure
CASE 2' Ref 1 constant, Ref 2 variable at the spots S=2...10
 'Db( )= drift threshold of Ref 2 according to Fig.2.10
  Db(1)=1.1:Db(2)=5.3:Db(3)=5.2:Db(4)=5:Db(5)=3.1
  Db(6)=3.7:Db(7)=-.1:Db(8)=-5.6:Db(9)=-12.8:Db(10)=-22.7
  FOR I=1 TO 10:Amp(I)=10^(Db(I)/20): NEXT ' amplitudes
  FOR T=1 TO 160:N(T)=But(T)^2' sum of energies
    IF T-T1>0 THEN N(T)=N(T)+(Amp(1)*But(T-T1))^2
    IF T-T(S)>0 THEN N(T)=N(T)+(Vari*Amp(S)*But(T-T(S)))^2
    N(T)= SQR(N(T)): NEXT ' here N(T) means sound pressure
CASE 3' Ref 1,2 constant, Ref 3 variable at the spots S=3...10
 'Db( )= drift threshold of Ref 3 according to Fig.2.10
  Db(1)=1.1:Db(2)=5.3:Db(3)=12:Db(4)=11:Db(5)=10
  Db(6)=8.2:Db(7)=4.5:Db(8)=.5:Db(9)=-8.5:Db(10)=-14.6
  FOR I=1 TO 10:Amp(I)=10^(Db(I)/20): NEXT ' amplitudes
  FOR T=1 TO 160:N(T)=But(T)^2' sum of energies
    IF T-T1>0 THEN N(T)=N(T)+(Amp(1)*But(T-T1))^2
    IF T-T2>0 THEN N(T)=N(T)+(Amp(2)*But(T-T2))^2
    IF T-T(S)>0 THEN N(T)=N(T)+(Vari*Amp(S)*But(T-T(S)))^2
    N(T)= SQR(N(T)): NEXT ' here N(T) means sound pressure
CASE 4' Ref 1,2,3 constant, Ref 4 variable at spots S=4...10
 'Db( )= drift threshold of Ref 4 according to Fig.2.10
  Db(1)=1.1:Db(2)=5.3:Db(3)=12:Db(4)=17.7:Db(5)=17.8
  Db(6)=16.5:Db(7)=15.9:Db(8)=10.7:Db(9)=1.8:Db(10)=-7.2
  FOR I=1 TO 10:Amp(I)=10^(Db(I)/20): NEXT ' amplitudes
  FOR T=1 TO 160:N(T)=But(T)^2' sum of energies
    IF T-T1>0 THEN N(T)=N(T)+(Amp(1)*But(T-T1))^2
    IF T-T2>0 THEN N(T)=N(T)+(Amp(2)*But(T-T2))^2
    IF T-T3>0 THEN N(T)=N(T)+(Amp(3)*But(T-T3))^2
    IF T-T(S)>0 THEN N(T)=N(T)+(Vari*Amp(S)*But(T-T(S)))^2
    N(T)= SQR(N(T)): NEXT ' here N(T) means sound pressure
CASE 5' Ref 1,2,3,4 const, Ref 5 variable at spots S=5...10
 'Db( )= drift threshold of Ref 5 according to Fig.2.10
  Db(1)=1.1:Db(2)=5.3:Db(3)=12:Db(4)=17.7:Db(5)=20.3
  Db(6)=22.3:Db(7)=20.1:Db(8)=18.5:Db(9)=10:Db(10)=-1.6
  FOR I=1 TO 10:Amp(I)=10^(Db(I)/20): NEXT ' amplitudes
  FOR T=1 TO 160:N(T)=But(T)^2' sum of energies
    IF T-T1>0 THEN N(T)=N(T)+(Amp(1)*But(T-T1))^2
    IF T-T2>0 THEN N(T)=N(T)+(Amp(2)*But(T-T2))^2
    IF T-T3>0 THEN N(T)=N(T)+(Amp(3)*But(T-T3))^2
    IF T-T4>0 THEN N(T)=N(T)+(Amp(4)*But(T-T4))^2
    IF T-T(S)>0 THEN N(T)=N(T)+(Vari*Amp(S)*But(T-T(S)))^2
    N(T)= SQR(N(T)): NEXT ' here N(T) means sound pressure
CASE 6' Ref 1,2,3,4,5 const, Ref 6 variable at spots S=6...10
 'Db( )= drift threshold of Ref 6 according to Fig.2.10
  Db(1)=1.1:Db(2)=5.3:Db(3)=12:Db(4)=17.7:Db(5)=20.3
  Db(6)=23.4:Db(7)=22.2:Db(8)=19.8:Db(9)=13.8:Db(10)=3.6
  FOR I=1 TO 10:Amp(I)=10^(Db(I)/20): NEXT ' amplitudes
  FOR T=1 TO 160:N(T)=But(T)^2' sum of energies
  IF T-T1>0 THEN N(T)=N(T)+(Amp(1)*But(T-T1))^2
  IF T-T2>0 THEN N(T)=N(T)+(Amp(2)*But(T-T2))^2
  IF T-T3>0 THEN N(T)=N(T)+(Amp(3)*But(T-T3))^2
  IF T-T4>0 THEN N(T)=N(T)+(Amp(4)*But(T-T4))^2
  IF T-T5>0 THEN N(T)=N(T)+(Amp(5)*But(T-T5))^2
  IF T-T(S)>0 THEN N(T)=N(T)+(Vari*Amp(S)*But(T-T(S)))^2
  N(T)= SQR(N(T)): NEXT ' here N(T) means sound pressure
END_SELECT
```

```
'b) ------------------------------------------ fast integrator RC=1.4ms
 FOR T=160 TO 1 STEP -1: FOR J=1 TO T-1
     N(T)=N(T)+N(T-J)*EXP(-J/1.4): NEXT J:N(T)=N(T)/1.4: NEXT T
'c) -------------------------------- loudness according to ISO standard
 FOR T=1 TO 160:N(T)=N(T)^.60206: NEXT ' ISO-exponent=2*LOG10(2)
'd) ----------------------- diode und RC-storage according to Vogel [6]
'Vogel Fig.9 RC=45ms, in the text 35ms. Niese 28ms with p^0.6
 E= EXP(-1/28): FOR T=1 TO 159:N(T+1)= MAX(N(T+1),N(T)*E): NEXT
'e) ------------------------------ final integrator, 3xRC=16ms low-pass
 FOR T=160 TO 1 STEP -1: FOR J=1 TO T-1
     N(T)=N(T)+N(T-J)* EXP(-J/16): NEXT J:N(T)=N(T)/16: NEXT T
 FOR T=160 TO 1 STEP -1: FOR J=1 TO T-1
     N(T)=N(T)+N(T-J)* EXP(-J/16): NEXT J:N(T)=N(T)/16: NEXT T
 FOR T=160 TO 1 STEP -1: FOR J=1 TO T-1
     N(T)=N(T)+N(T-J)* EXP(-J/16): NEXT J:N(T)=N(T)/16: NEXT T
'------------------ loudness N(T) ready, store curve ------------------
 IF Vari=0 THEN FOR T=1 TO 160:Nzero(T)=N(T): NEXT' basic curve
 IF Vari>0 THEN FOR T=1 TO 160:Nplus(T)=N(T): NEXT' + reflection
'----------------------------------------------------------------------
 IF Vari=0 THEN :Clear5: PRINT @(5,8);
    PRINT "s= screen c= print curves a= print angles";
    PRINT @(6,0);"Output: ";: REPEAT Out$= LOWER$( INPUT(1))
    UNTIL Out$="s" OR Out$="c" OR Out$="a" ENDIF
 IF Out$="c" THEN :Clear5' ------------------------- curves to printer
    LPRINT "!R!; res; daf; unit c; rel 600;"' Kyocera Prescr. II
    LPRINT "sls .4; fset 0p";15;"h0s0b4099T;"'4mm lines, Courier
    LPRINT "text 'date "+ DATE$ +" clock "+ TIME$ +"',N;"
    LPRINT "text 'Dir sound";Db0-3;"dB, "
    IF Vari=0 THEN : LPRINT "basic curve for Ref";Ref;"',N;"
    ELSE LPRINT "Reflection";Ref;" at spot";S;"',N;" ENDIF
    LPRINT "text 'ms    N    Npoint',N;"
    FOR T=1 TO 151 STEP 3: LPRINT "text '";: FOR J=0 TO 2
        LPRINT T+J;" ";N(T+J);" ";N(T+J+1)-N(T+J);" ";
        NEXT J: LPRINT "',N;": NEXT T: LPRINT "page; exit;"
    IF Vari=0 THEN :Vari=1: GOTO Again ENDIF ENDIF
 IF Out$="a" THEN :Clear5' ------------------------- angles to printer
    IF Vari=0 THEN ' Kyocera, language Prescribe II
    LPRINT "!R!; res; daf; unit c; rel 600;"' initialize
    LPRINT "sls .4; fset 0p";15;"h0s0b4099T;"' lines, font
    LPRINT "text 'date "+ DATE$ +" clock "+ TIME$ +"',N;"
    LPRINT "text 'Dir sound";Db0-3;"dB, "
    LPRINT "basic curve Nzero for Ref";Ref;"',N;"
    LPRINT "text 'ms Nzero Npoint',N;"
    FOR T=1 TO 151 STEP 3: LPRINT "text '";: FOR J=0 TO 2
    LPRINT T+J;" ";N(T+J);" ";N(T+J+1)-N(T+J);" ";
    Next J: LPRINT "',N;": NEXT T
    LPRINT "text ' T+23ms Nplus(T+23) ";
    LPRINT "Nptzero__D__Nptplus__D__Tangdelta__D__"
    LPRINT "in_scale_diagram',N;"
 ELSE ' ------------------------------------------- deviation angle
   'angle graphically correct only with:                  10ms<-->1sone
    Nptzero=(Nzero(T(S)+24)-Nzero(T(S)+22))*5'            =(diff/2ms)*10
    Nptplus=(Nplus(T(S)+24)-Nplus(T(S)+22))*5'            =(diff/2ms)*10
    Tangdelta!=(Nptplus-Nptzero)/(1+Nptplus*Nptzero)
    LPRINT "text 'spot ";S;" ";T(S);"+23ms ";
    LPRINT Nplus(T(S)+23);" ";Nptzero;" ";Nptplus;" ";
    LPRINT Tangdelta!;"',N;" ENDIF
 IF Vari=0 THEN Vari=1: GOTO Again
 IF S=10 THEN: LPRINT "page; exit;" ENDIF ENDIF
```

```
IF Out$="s" THEN ' ---------------------------------------- use screen
   IF Vari=0 THEN : USING ' -------------------------------- curve Nbas
      Clear5: PRINT @(5,0);"date "+ DATE$ +" clock "+ TIME$
      PRINT @(20,3);"Basic curve:";
      PRINT @(22,0);" T Nzero Npoint";
      FOR T=1 TO 151 STEP 3: PRINT @(23,0); CHR$(27);"J";
      FOR J=0 TO 2: PRINT @(23,J*25); USING "####";T+J;
          PRINT USING "###.####";N(T+J);N(T+J+1)-N(T+J);
          NEXT J: INPUT @(23,75);Pause: NEXT T
      Vari=1: GOTO Again ENDIF USING
     '---------------------------------------------- deviation angle
     'angle graphically correct only with:              10ms<-->1sone
      Nptzero=(Nzero(T(S)+24)-Nzero(T(S)+22))*5'      =(diff/2ms)*10
      Nptplus=(Nplus(T(S)+24)-Nplus(T(S)+22))*5'      =(diff/2ms)*10
      Tangdelta!=(Nptplus-Nptzero)/(1+Nptplus*Nptzero)
      PRINT @(20,2);"Spot";S;" =";T(S);"+23ms ";
      PRINT @(22,0);" N    Nptzero_Nptplus____Tangdelta";
      PRINT "___in_scale_diagram";
      PRINT @(23,0); SPC(70);@(23,0);: USING "###.####"
      PRINT N(T(S)+23);" ";Nptzero;Nptplus;" ";Tangdelta!;
      REPEAT INPUT @(23,36);Pause: WAIT .1
      UNTIL WPEEK($FFFC02)>0: GOTO Again ENDIF NEXT S: STOP
DEF PROC Clear5: PRINT @(5,0); CHR$(27);"J";@(5,0);: END_PROC
DATA .12,.215,.29,.36,.425,.48,.53,.575,.62,.663,.705'            1...11
DATA .74,.775,.807,.84,.871,.903,.931,.956,.974,.986'               ...21
DATA .994,1.,1.,1.001,1.001,1.003,1.003,1.004,1.005,1.005'          ...31
DATA 1.01,1.01,1.01,1.01,1.01,1.005,1.,.99,.98,.965,.95'            ...42
DATA .93,.91,.89,.86,.83,.79,.75,.7,.65,.59,.53,.47,.41'            ...55
DATA .36,.31,.28,.26,.245,.235,.225,.215,.205,.195,.185'            ...66
DATA .175,.17,.16,.155,.15,.14,.135,.125,.12,.12,.11,.11'           ...78
DATA .1,.1,.09,.09,.08,.08,.08,.07,.07,.06,.06,.05,.05'             ...91
DATA .05,.04,.04,.04,.03,.03,.03,.02,.02,.01,.01,.01: END
```

References

1. W. Burgtorf, Acustica **11**, 97 (1961)
2. H. Haas, Einfluss eines Einfachechos auf die Hörsamkeit von Sprache. Acustica **1**, 49 (1951)
3. E. Meyer, G.R. Schodder, Nachr. Akad. Wissensch. Göttingen. Math.-Phys. Kl. **6**, 31 (1952)
4. R. Thiele, Richtungsverteilung und Zeitfolge der Schallrückwürfe in Räumen. Acustica **3**, 291 (1953)
5. E. Meyer, R. Thiele, Raumakustische Untersuchungen in zahlreichen Konzertsälen und Rundfunkstudios unter Anwendung neuerer Messverfahren. Acustica **6**, 431 (1956)
6. A. Vogel, Ein gemeinsames Funktionsschema zur Beschreibung der Lautheit und der Rauhigkeit. Biol. Cybernetics **18**, 31 (1975)
7. E. Zwicker, A model describing temporal effects in loudness and threshold, in *The 6th Intern. Congr. on Acoustics*, Tokyo, 1968, paper A-3-4
8. R. Strauss, Till Eulenspiegels lustige Streiche (Philips, 1981), Compact Disc 411 442-2
9. E. Meyer, W. Burgtorf, P. Damaske, Eine Apparatur zur elektroakustischen Nachbildung von Schallfeldern. Subjektive Hörwirkungen beim Übergang Kohärenz-Inkohärenz. Acustica **15**, 339 (1965)
10. M.R. Schroeder, D. Gottlob, K.F. Siebrasse, Comparative study of European concert halls. J. Acoust. Soc. Am. **56**, 1195 (1974)
11. M. Sakurai, Y. Korenaga, Y. Ando, *Music and Concert Hall Acoustics* (Academic Press, London, 1997), p. 51
12. J. Jecklin, *Musikaufnahmen* (Franzis, München, 1984)
13. Denon Professional Test CDs, Compact Discs PG-6013→15
14. Atari computer Falcon030, IDE hard disk 800 MB, MO disk 500 MB, 16 bit audio processing. Programming language Omikron-Basic. Analog circuits on main board trimmed for precision
15. S. Mehrgardt, V. Mellert, Transformation characteristics of the external human ear. J. Acoust. Soc. Am. **61**, 1567 (1977)
16. P. Damaske, B. Wagner, Richtungshörversuche über einen nachgebildeten Kopf. Acustica **21**, 30 (1969)

17. E.A.G. Shaw, R. Teranishi, J. Acoust. Soc. Am. **44**, 240, 257 (1968),
18. J. Blauert, *Räumliches Hören* (Hirzel, Stuttgart, 1974)
19. B.S. Atal, M.R. Schroeder, Gravesaner Blätter **27/28**, 124 (1966)
20. P. Damaske, V. Mellert, Ein Verfahren zur richtungstreuen Schallabbildung des oberen Halbraumes über zwei Lautsprecher. Acustica **22**, 153 (1969/70)
21. P. Damaske, Head-related two-channel stereophony with loudspeaker reproduction. J. Acoust. Soc. Am. **50**, 1109 (1971)
22. Probe microphones MD 321, Sennheiser
23. A.W. Mills, J. Acoust. Soc. Am. **30**, 237 (1958)
24. Condensor microphones MKH 106, Sennheiser
25. Analog tape recorder Revox A700, tape speed 38 cm/s, noise-reduced tape Ampex type 456 in main experiment, frequency response and channel symmetry carefully controlled
26. Condensor microphones MKH 40 P 48, Sennheiser. Measuring instrument Unigor 6e, 1.5% acccuracy up to 20 kHz, Goerz Electro
27. U. Tietze, Ch. Schenk, *Halbleiter-Schaltungstechnik* (Springer, Berlin, 1974)
28. P. Damaske, Richtungsabhängigkeit von Spektrum und Korrelationsfunktionen der an den Ohren empfangenen Signale. Acustica **22**, 191 (1969/70)
29. H. Bonsel, *Kuppler ("künstliches Ohr") für die Messung von Hörgeräten*. Grüne Ausbildungsmappe, vol. 12 (Median-Verlag, Heidelberg, 1975)
30. Equalizer SH-8075, Technics
31. L. Cremer, *Vorlesungen über technische Akustik* (Springer, Berlin, 1971), p. 315
32. P. Damaske, Subjektive Untersuchung von Schallfeldern. Acustica **19**, 199 (1967/68)
33. Y. Ando, *Concert Hall Acoustics* (Springer, Berlin, 1985)
34. E. Meyer, E. Neumann, *Physikalische u. technische Akustik* (Vieweg, Braunschweig, 1967), p. 79
35. L. Cremer, *Statistische Raumakustik* (Hirzel, Stuttgart, 1961), p. 120
36. M.R. Schroeder, *Number Theory in Science and Communication* (Springer, Berlin, 1997)
37. R. Feldtkeller, E. Zwicker, *Das Ohr als Nachrichtenempfänger* (Hirzel, Stuttgart, 1956)
38. Y. Ando, D. Noson, *Music and Concert Hall Acoustics* (Academic Press, London, 1997), p. 171
39. H. Niese, Hochfrequenztechnik u. Elektroakustik **70**, 5 (1961)
40. W. Kraak, H. Weißing, *Schallpegelmeßtechnik* (VEB Technik, Berlin, 1970)
41. G. Boré, Kurzton-Meßverfahren zur punktweisen Ermittlung der Sprachverständlichkeit in lautsprecherbeschallten Räumen. Dissertation, Technische Hochschule Aachen, 1956
42. J.P.A. Lochner, J.F. Burger, Acustica **11**, 195 (1961)
43. L.L. Beranek, *Music, Acoustics and Architecture* (Wiley, New York, 1962)
44. H. Kuttruff, *Room Acoustics* (Elsevier, London, 1991)
45. J.C.R. Licklider, A duplex theory of pitch perception. Experientia **7**, 128 (1951)
46. J.C.R. Licklider, A triplex theory of pitch perception, in *3rd London Symposium on Information Theory*, ed. by C. Cherry (Butterworth's, London, 1956)
47. Y. Ando, *Architectural Acoustics* (Springer, New York, 1998)
48. B.S. Atal, H. Kuttruff, M.R. Schroeder, in *Proceedings of the Fourth International Congress on Acoustics*, Copenhagen, 1962, paper H31

Index

Printed in the United States
138823LV00002B/31/P

9 783540 782278